The Institute of Biol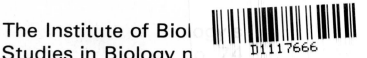
Studies in Biology n

Decomposition

C. F. Mason

B.Sc., D.Phil.
University of East Anglia
now at Water Data Unit, Department of the Environment, Reading

Edward Arnold

First published 1977
by Edward Arnold (Publishers) Limited,
25 Hill Street, London W1X 8LL

Board edition ISBN: 0 7131 2588 8
Paper edition ISBN: 0 7131 2589 6

Printed in Great Britain by
The Camelot Press Ltd, Southampton

General Preface to the Series

It is no longer possible for one textbook to cover the whole field of Biology and to remain sufficiently up to date. At the same time teachers and students at school, college or university need to keep abreast of recent trends and know where the most significant developments are taking place.

To meet the need for this progressive approach the Institute of Biology has for some years sponsored this series of booklets dealing with subjects specially selected by a panel of editors. The enthusiastic acceptance of the series by teachers and students at school, college and university shows the usefulness of the books in providing a clear and up-to-date coverage of topics, particularly in areas of research and changing views.

Among features of the series are the attention given to methods, the inclusion of a selected list of books for further reading and, wherever possible, suggestions for practical work.

Reader's comments will be welcomed by the author or the Education Officer of the Institute.

1976

The Institute of Biology,
41 Queen's Gate,
London, SW7 5HU.

Preface

Plants and animals require a constant source of nutrients for growth and the death and decomposition of organisms, with release of nutrients, is an essential link in the maintenance of nutrient cycles.

I have attempted here to look at some of the details of the subtle interplay of abiotic factors, the microflora and the fauna in the decomposition process as well as placing decomposition in a wider ecological perspective. Some important applied aspects of the process are also described.

Reading, 1976

C. F. M.

'Death is not in the nature of things; it is the nature of things.
But what dies is the form. The matter is immortal.'

John Fowles

Contents

1 Introduction

1.1 Processes within ecosystems

The flow of energy and the cycling of nutrients are two of the major processes taking place within ecosystems. They are illustrated very simply in Fig. 1-1. In the energy flow diagram (Fig. 1-1a) the boxes represent

(a)

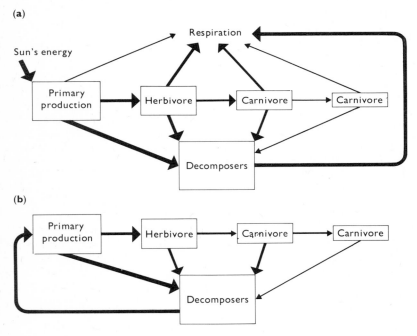

(b)

Fig. 1-1   (a) The flow of energy and (b) the cycling of nutrients, through an ecosystem.

standing crops or *biomass* and the arrows fluxes. The green plants (*autotrophs* or *primary producers*) utilize the energy of the sun to fix carbon dioxide and produce energy-rich organic compounds. A proportion of this production is lost to the ecosystem as respiration as the green plants utilize organic compounds to do work. A variable, but usually small, proportion of this production is eaten by herbivores (*heterotrophs* or *primary consumers*) and the remainder eventually senesces and enters the decomposer system. The herbivores also dissipate energy in respiration.

The faeces they produce go directly to the decomposer chain. Some herbivores are eaten by carnivores, others die and are decomposed. The flow of energy through the carnivores is similar. All the energy which is not dissipated as heat of respiration enters the decomposer system, where it is utilized for work. In a steady state ecosystem the energy fixed by the autotrophs is equal to that lost by the ecosystem as respiration. Energy flow through ecosystems is described in detail by PHILLIPSON (1966).

During the energy transfers there are also movements of nutrients. In Fig. 1–1b the boxes are now usually termed *pools* and the arrows again are fluxes. All nutrients ultimately enter the decomposer system where the breakdown of organic material releases them in a form such that they can readily be taken up again by the vegetation. The continued circulation of nutrients is essential for the growth of plants.

It is obvious, then, that decomposition is an important process in the functioning of ecosystems.

1.2   Definitions of decomposition

The Oxford English Dictionary defines the verb to decompose as 'to separate or resolve into its constituent parts or elements'. Dead organisms are broken down into large particles, then into smaller particles and eventually into small molecules. The dead organism is thus gradually disintegrated until its structure can no longer be recognized and complex organic molecules are broken down into carbon dioxide, water and mineral components. Decomposition involves a complex interaction of physical and biological agencies. The organisms involved are known as *decomposers*, *reducers* or *saprobes*.

A slight distinction should be made here between decomposition and breakdown. The loss of weight which is often measured in experiments (for instance of material contained within litter bags) is largely the breakdown of tissues from large to small particles and is thus only the initial part of the decomposition process as defined above.

2 The Production of Litter

2.1 Introduction

Organic detritus or litter can be defined as 'all types of biogenic material in various stages of decomposition which represent potential energy sources for consumer species' (DARNELL, 1967). The term litter is usually used in terrestrial systems and especially for material derived from higher plants. Detritus is in more general usage in studies of aquatic ecosystems. The majority of litter is derived from plant sources, though that from animals is sometimes considerable, for instance the dung produced by populations of large herbivores.

2.2 Litter fall in forests

The plant material entering the litter layer of a forest soil is a diverse mixture of the various components of the plants' structure. Although in temperate areas most of the litter falls in autumn, amounts produced at other times of the year should not be ignored.

To illustrate the structural and temporal patterns of litter fall, the results of a study in a stand of beech trees (*Fagus sylvatica*) in Wytham Woods, near Oxford (MASON, 1970b), are shown in Fig. 2–1. The trees were about two hundred years old and the fall of litter was measured for a year. Litter was collected in conical bags made of sail-cloth, with a diameter of 1 m and placed 1 m above the ground. Twenty such litter bags were sited randomly beneath the trees and their contents were emptied at the beginning of each month (see Fig. 2–2). The litter was divided into its various components and dried to constant weight.

The majority (72%) of the beech litter fell in the period October–December and was composed of leaves and, to a lesser extent, fruits. Bud-scales and flowers made up only 5.2% of the total litter fall, but they were a very important source of input to the litter layer in May. Twigs and bark made up 10% of the total and the highest fall was in September, during a period of gales, but it was relatively constant over the rest of the year. No account was taken in this study of large branches whose fall is erratic and localized, but over a period of years these are very important.

The total litter fall on the Oxford site was 652 g/m²/annum (6.52 tonnes/hectare/annum) of which 584 g/m² was beech material, the remainder being mainly hornbeam and sweet chestnut from the edge of

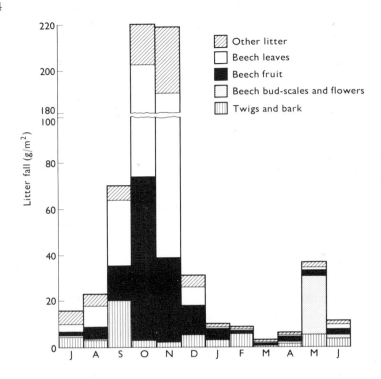

Fig. 2–1   The pattern of litter fall throughout the year in a stand of beech (*Fagus sylvatica*), Wytham Wood, near Oxford.

the site. Only 58% of the litter fall was leaves so it is clearly of little use to measure just the input of leaves in order to gain an understanding of the production, decomposition and nutrient cycling processes occurring within forests.

Similar studies of litter-fall under different canopies are relatively easy to perform and polythene buckets make perfectly adequate litter collectors.

As well as the trees, the shrub and herb layers may add significant amounts of litter to the forest floor. Beech woodland produces heavy shade so the field layer is sparse and a potential litter input of only 23.3 g/m²/annum of litter (mainly dog's mercury, *Mercurialis perennis*) was estimated for the Oxford stand. In woodlands with a more open canopy, for instance oak or sycamore, the amount of litter produced by the shrub and field flora is likely to be much higher.

The different fractions of the litter have very different structures and

Fig. 2–2   Litter bags within a stand of beech trees, Wytham Wood, near Oxford. (Photo. J. R. Flowerdew.)

chemical composition and will decompose at different rates. The overall decomposition rate of the litter will therefore depend on the relative proportions of the different components.

Temperate deciduous forests show a marked seasonal peak in litter fall with most occurring in the autumn. Some conifers in temperate forests, for instance Scots pine, *Pinus sylvestris*, also show a seasonal pattern of litter fall, whereas others, for instance spruce, *Picea abies*, do not. Litter fall tends to be continuous throughout the year in humid, equatorial forests, whilst in a dry, tropical region in West Africa *Acacia alba* sheds leaves mainly in the rainy season (JUNG, 1969). A seasonality in litter input is likely to have a marked effect on the biology of decomposer organisms.

Different tree species produce different amounts of litter. The litter fall of a representative sample of species from various climatic regions is shown in Table 1. The total litter fall ranges from 1.5 t/ha/annum for spruce in Norway to 23.3 t/ha for a tropical rain-forest in South-east Asia. Leaves form a variable proportion of the total litter fall, ranging from 21.7% for *Acacia* forest in West Africa to 80.7% for spruce in Norway.

A number of factors influence the quantity of litter produced by forests but climate is undoubtedly the most important. This is shown in a general

Table 1 The annual amounts of litter-fall of various tree species.

Dominant forests		Locality	Litter Fall (t/ha/annum)		% Leaves	Reference
			Leaf litter	Total litter		
ALPINE AND ARCTIC FORESTS						
Spruce	*Picea abies*	Norway	0.9	1.5	60.0	MORK, 1942
COOL TEMPERATE FORESTS						
	Picea abies	Norway	2.5	3.1	80.7	MORK, 1942
Pine	*Pinus sylvestris*	Finland	1.7	2.3	73.9	VIRO, 1955
Beech	*Fagus sylvatica*	England	3.4	5.8	58.6	MASON, 1970b
		South Sweden	3.6	5.7	63.2	ANDERSSON, 1970
Oak	*Quercus petraea*	England	2.1	3.9	53.9	CARLISLE *et al.*, 1966
	Quercus robur	South Sweden	3.3	5.3	62.3	ANDERSSON, 1970
WARM TEMPERATE FORESTS						
	Eucalyptus diversicolor	Western Australia	2.7–2.8	5.7–6.0	47.0	STOATE, 1958
Evergreen oak	*Quercus ilex*	Southern France	2.5–2.7	3.8–7.0	38.6–65.7	RAPP, 1969
TROPICAL FORESTS						
	Acacia alba	West Africa	2.5	11.5	21.7	JUNG, 1969
	Diospyros spp.	Ghana	7.0	10.5	66.7	NYE, 1961
Tropical rain forest		Thailand	11.9	23.3	51.1	KIRA and SHIDEI, 1967

way by examining Table 1, where marked differences in litter production with latitude are apparent. BRAY and GORHAM (1962) have shown that the ratios of litter production are 1 : 3 : 5 : 10 for the Arctic-Alpine, Cool Temperate, Warm Temperate, and Equatorial climatic zones respectively, with a mean annual temperature ranging from less than zero to 25°C. The length of time that temperatures are above freezing is also important as is the amount of sunlight. The growing period for trees in northern Scandinavia may be as little as three months, whereas in cool temperate regions it is in excess of six months. A similar climatic effect is seen with altitude. MORK (1942) studied birch (*Betula*) forests at 180 m and 800 m above sea level and found an annual litter fall of 1.88 and 0.80 t/ha respectively.

Climate also has a marked effect on litter fall between years. In the relatively stable climates of equatorial regions there is little annual variation in litter fall, but it may vary as much as two times in temperate deciduous Angiosperm forests and as much as five times in Gymnosperm forests in similar climates.

PHILLIPSON *et al.* (1975) studied the production of beech litter over a period of four years in a temperate woodland near Oxford. The litter fall varied between 205 and 388 g/m²/annum, but the production of non-wood litter (leaves, flowers and bud scales) was remarkably constant, with a mean value of 148 g/m²/annum. Twig production (52 g/m²/annum) also varied little. Most of the difference between years was due to the production of fruit, beech mast composing from 2% to 54% of the total tree litter fall. Fruit production in temperate forest trees is notoriously variable from year to year but heavy mast years are often synchronous over wide geographical areas.

In contrast to climate, the soil conditions have little influence on the quantity of litter produced. Little difference was found between beech growing on mull and mor soils, but conifers generally produce more litter on fertile soils, with trees on the poorest sites producing only a third as much litter as trees on good quality soils.

The age of the trees is important in the early years, with the amount of litter produced increasing with age until the canopy is closed. The thinning of the canopy causes a decrease in litter production and the continuous removal of litter from the forest floor eventually reduces the annual litter input, presumably because the nutrient pool within the soil becomes impoverished.

2.3   Roots

The above discussion of litter fall made no mention of dead root tissue and this has received little attention, chiefly because of the difficulties encountered in sampling and in determining whether a section of root is alive or dead.

The primary functions of the root system are to provide anchorage and to take up water and nutrients from the soil. Those parts of a plant which are below ground are probably as extensive as the aerial parts. The root systems of forest trees are often heavier than the canopies and the root system of reeds *Phragmites communis* account for up to 40% of the total biomass of the plant, even when the shoots are fully grown. The annual production of beech roots has been estimated at 62% of the leaf production and PHILLIPSON *et al.* (1975) used this value to calculate a root production of 92 g/m²/annum on their study area (see above).

Many of the small lateral roots of a plant live only for a short while. Lateral roots also tend to lose their epidermis and cortex in warm soils, with up to half the dry weight being shed in a year. A large amount of root tissue would appear then to be available for decomposition.

2.4   Other terrestrial habitats

These will range from dry desert areas with sparse vegetation, through savannas, temperate grasslands and agricultural land to moorland and tundra. Their biomass, and hence litter production, is much lower than that of forests. The biomass of desert vegetation is of the order of 7 t/ha, savanna 40 t/ha and temperate grasslands 15 t/ha. The tundra biomass is similar to that of deserts. The litter production in these areas has received little study but many of the habitats are maintained by burning, grazing or cropping and there is no build-up of litter. For instance, once the grazing pressure is removed from a grassland of *Brachypodium pinnatum*, it will quickly develop a thick litter layer and tussocky appearance. An associated rich flora, adapted to a regime of heavy grazing, is overgrown and disappears. This has happened in many areas of Britain since myxomatosis decimated the population of rabbits.

2.5   Freshwater habitats

Material from other ecosystems is often an important source of litter or detritus in freshwater habitats. Such material is described as *allochthonous*; that produced within the ecosystem is termed *autochthonous*. Allochthonous litter is especially important where the edge of the aquatic habitat is large relative to its surface area, notably in rivers, streams and ponds. The surrounding trees are responsible for the majority of the energy input in the form of autumnal leaves. The currents of running waters ensure that phytoplankton populations are usually low, but attached algae (the *periphyton*) can be a significant source of nutrition to grazing animals and detritivores. Even when abundant the yearly production of higher plants in running waters is only of the order of 1–3 t/ha and they seem more important as sites of attachment and shelter for other organisms rather than as a source of food.

Autochthonous production is of greater significance in lakes, where the surface area relative to the edge is larger. Phytoplankton is the chief producer, with macrophytes usually confined to the littoral zone. Phytoplankton populations tend to be subject to heavy grazing pressures so a smaller proportion of their production will enter the detritus food chain directly than is the case with most other primary producers. The benthic fauna of lakes is dependent on the decomposition cycle for most of its food.

Nutrient-rich lakes are skirted by reedswamps which are often extensive. Reedswamps are amongst the most productive habitats in the world. A reedswamp in the Norfolk Broads had an average annual production of reed, *Phragmites communis*, of 8.2 t/ha and of reedmace, *Typha angustifolia*, of 17 t/ha (MASON and BRYANT, 1975). All the above-ground parts die down in winter and they are subject to negligible grazing so potentially a large amount of detritus will enter aquatic habitats from reedswamps. The litter production of reedswamps is substantially higher than for temperate forests, as can be seen from the data in Table 1.

2.6   Marine habitats

DARNELL (1967) has examined the sources of detritus in an estuarine ecosystem. The autochthonous sources included the phytoplankton, marginal submerged vegetation, the algae growing on the mud-flat and the periphyton growing on emergent plants. The allochthonous sources include marginal marsh and swamp vegetation, as well as plankton and detritus originating from the rivers. Material from the shore, brought in during periods of high water, and windblown detritus, can also be significant. There are also phytoplankton and detritus brought in from the adjacent marine habitats. The relative importance of these sources varies from estuary to estuary as well as seasonally. In an estuary studied by ODUM and DE LA CRUZ (1967), 95% of the detritus was derived from the salt marsh grass *Spartina*, with very little from phytoplankton and animals.

The litter present on the coasts and in the open sea is also of very variable composition. The larger seaweeds are often an important component. MANN (1972), for instance, has shown that laminarians in Nova Scotia produce an average of 1750 g/m^2 carbon annually, of which 90% enters the detritus pool. Periodic strandings of crustacean exuviae, hydroids, etc., give an occasional increased animal element to the litter. Wind and tidal currents have a marked effect on the movement and distribution of coastal litter.

The benthic fauna of the oceans, as with lakes, is dependent on the rain of plants and animals from the photic zone of the surface layers of the sea for food.

2.7   Animal litter

The input of animal litter to the detritus food chain will consist chiefly of corpses and faeces. Exuviae, hair, the pellets produced by carnivorous birds, etc., will normally be of very minor importance, though they may have very specialized decomposer organisms associated with them. Corpses are of only localized occurrence under normal conditions but they provide a rich protein food source. Animal dung normally forms only a small proportion of the total detrital input. It will, however, assume greater importance when for instance, plagues of caterpillars are defoliating forests, especially as faeces are rapidly colonized and broken down, thus speeding nutrient cycling. Dung is also a significant component in rangelands supporting high populations of large herbivores. A single cow will produce some 8 t of dung each year. COE (1972) estimated that the elephant population on the savannas of Tsavo National Park consumed 22% of the primary production and produced 154 kg (wet weight) dung/km^2 each day.

2.8   Man-made litter

Very large amounts of waste are produced by the human population and the disposal of this is of great concern. The amounts of some organic wastes produced in Britain are shown in Table 2. Some of these, of course, can be re-used, though it is often not economic to do so. For instance in earlier years, farm manure was spread on fields as fertilizer, but with today's intensive animal farming units on small areas of land, together with high labour costs, this is often no longer practicable (though the recent soaring prices of artificial fertilizers may make it more economic in future). The release of sewage sludge (including phosphate-rich detergent waste) and farm manure into rivers and lakes has had a very damaging effect on the natural communities by increasing the nutrient loading and

Table 2   The production of waste in Britain. (From GRAY, K. R., SHERMAN, K. and BIDDLESTONE, A. J., 1971.)

Material	Tonnes per year, fresh weight
Wood-shavings and sawdust	1 070 000
Straws—wheat, barley, oats	3 000 000
Potato and pea haulms, sugar-beet tops	1 600 000
Garden wastes	1 000 000—10 000 000
Sewage sludge	20 000 000
Municipal refuse	18 000 000
Farm manures	120 000 000

decreasing the oxygen (see for example MELLANBY, 1972). Urban rubbish is also a problem because large areas of land are required for tipping and many of the waste components are not readily broken down by decomposer organisms.

3 The Basic Process

3.1 Introduction

The previous chapter emphasized the very diverse origin and structure of materials available for decomposition and it will be obvious that this implies a considerable biochemical diversity. The majority of litter consists of the structural components of plants, such as cellulose, hemicellulose and lignins, all of which are broken down relatively slowly.

3.2 Biochemistry of decomposition

Cellulose is the major structural constituent of plant cells, making up, for instance, 90% of cotton fibres and 50–60% of wood. Different molecular weights for cellulose are reported from different sources, that from spruce having a molecular weight of 220 000 while cotton fibre cellulose has a molecular weight of 330 000. The molecules contain between 1400 and 10 000 glucose residues. Cellulose molecules have a thread-like structure and are combined in micelles of some sixty molecules. These micelles are formed by hydrogen-bonding between the hydroxyl groups of cellulose and the molecules of water which are adsorbed by the cellulose. A large number of bonds make the micelle very stable. The cellulose molecule has the following structure:

The cellobiose residues are linked into the long chain by glycosidic bonds. The first stage in the enzymic breakdown of cellulose involves the formation of linear chains. A second enzyme, a 1,4,β-gluconase, then hydrolyzes these chains to produce molecules of cellobiose, which are then hydrolyzed, with the aid of β-glucosidase, to glucose:

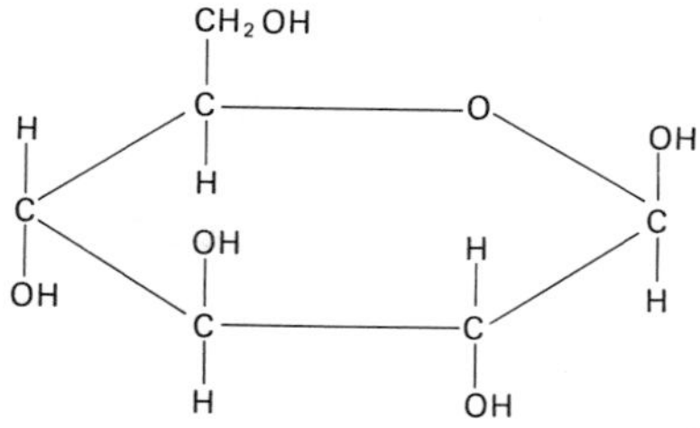

Hemicelluloses occur in smaller amounts than cellulose, making up, for instance, 12% of pine needles and 10–30% of wood. They are polysaccharides of high molecular weight, constructed from arabinose, galactose, mannose, xylose and uronic acids. Arabin, as an example, has the structure:

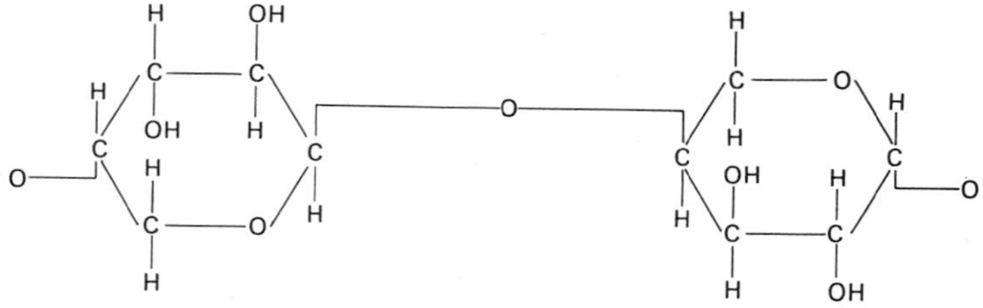

It differs from cellulose in that the CH_2O groups are replaced by hydrogen atoms. Other hemicelluloses have very different structures and the enzymes involved in their hydrolysis are very diverse and relatively unknown.

Lignin forms some 20–30% of wood and has a considerably more complex structure than cellulose or hemicelluloses. It appears to consist mainly of building blocks of phenyl propanes, such as coniferyl alcohol, with the benzene ring carrying a hydroxyl group and one or two methoxy groups. Phenyl propane units are joined together by both ethereal linkages and carbon–carbon bonds, these latter being highly resistant to chemical degradation. Phenyl propane units are joined in a number of ways, for example:

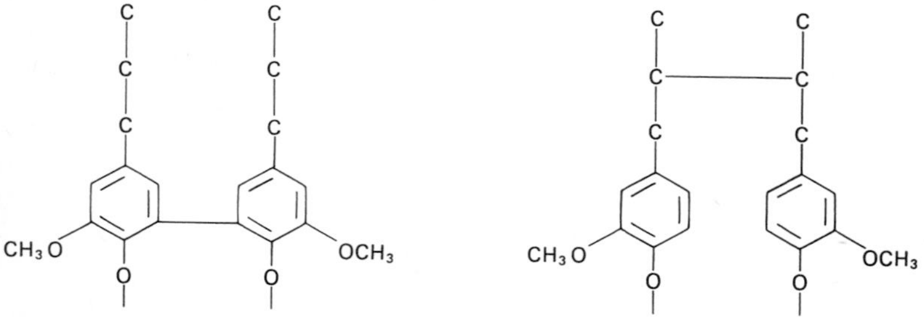

There is a good deal of evidence that all these linkages may occur in lignin, although there is as yet no rigorous proof.

The initial attack on lignin by micro-organisms, at least by white-rot fungi (see Chapter 4) results in the loss of methoxy groups and breaking of other ether linkages of the lignin, causing a partial degradation to more soluble substances. Examples of these intermediate products are guaiacylglycerol-β-coniferyl ether (a) and syringaldehyde (b):

(a)

(b)

The guaiacylglycerol units are then converted to vanillin or vanillic acid, among other compounds. These more soluble end products are probably then available for attack by a wider variety of organisms than were the initial molecules. It must be emphasized that the structure of lignin is still very inadequately understood, as is the biochemistry of its decomposition.

3.3　Stages of decomposition

Decomposition can be divided into three basic processes, leaching, weathering and biological action. They take place, of course, simultaneously. Leaching involves the rapid loss of soluble material from the detritus by the action of rain or water flow. Weathering is the mechanical breakdown of the detritus due to physical factors such as abrasion by wind, ice or the action of waves. Biological action results in the gradual fragmenting and oxidation of detritus by living organisms and is dealt with in detail in Chapters 4 and 5.

Various stages of decay of litter can be recognized, which can be examined by cutting a section through the litter layer of a woodland floor. MILLER (1974) described the stages in decay in the litter layer of a coniferous forest in the following way:

AoooL　freshly fallen, undecomposed needles

AooF　dark brown, intact, recognizable needles, extensively colonized by fungi

F_2　greyish, fragmented, compressed but recognizable needles containing hyphal fragments and animal faeces. The leaf mesophyll is collapsed

AoH　humus-like, amorphous mass of animal faeces, needles and microbial fragments

A_1　an intimate mixture of humus and mineral soil

An accumulation of dead material is fragmented and chemically graded by a variety of physical and biotic agents to produce an amorphous aggregate of chemically resistant materials, known as *humus*. Humus is a relatively stable complex and is broken down to its constituents only very slowly.

3.4　Leaching and weathering

Nutrients such as carbohydrates and amino acids, are largely withdrawn from senescing parts of a plant into the storage organs. At the same time waste products are translocated into the dying tissues so that senescent and functional components differ chemically.

Decomposing material contains many soluble compounds which are removed rapidly by the abiotic process of leaching. Leaching occurs throughout the life of a plant due to the effects of rainfall and substantial amounts of nutrients enter the soil of a forest from the canopy during the growing season of trees. Losses of potassium, for instance, are of the order of 10–200 kg/ha/year and nitrogen 1–20 kg/ha/year, the higher losses occurring in tropical rainforests subject to heavy rainfall. Substantial quantities of carbohydrates are also lost.

Weight and nutrient losses from litter due to leaching occur most rapidly in the first month of entry into the litter layer. BOYD (1970) placed dried litter of *Typha latifolia* in fibreglass bags. One group of bags was suspended 12 cm above water, the other group was submerged 30 cm below the surface in water and the weight and nutrient losses were studied. The results are shown in Fig. 3–1. The weight loss of litter was

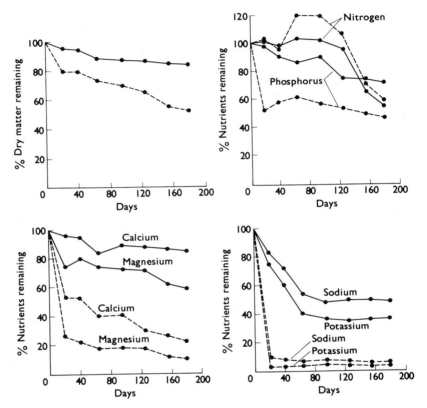

Fig. 3–1   Losses of dry matter and nutrients from litter of *Typha latifolia* (reedmace) in fibreglass bags suspended 12 cm above water surface (solid line) and submerged 30 cm below water surface (dashed line). (Reproduced with permission from BOYD, (1970.)

much greater in the submerged bags than in the suspended bags, especially during the first twenty days, when much leaching occurred. Sodium and potassium rapidly entered soluble states and were leached out, while the loss of calcium and magnesium was less rapid, with more remaining at the end of the experiment. Similarly, phosphorus had a greater residual level. In a study of the decomposition of *Typha angustifolia* and *Phragmites communis* in Alderfen Broad, Norfolk (MASON and BRYANT, 1975), 76% of sodium, 93% of potassium, 70% of calcium and 65% of magnesium were leached out of the litter in the first month. Calcium and magnesium are more intimately bound up within complex structural molecules in the plants than are sodium and potassium, so they leached out less readily.

From Fig. 3–1 it can be seen that there is an increase in nitrogen in *Typha latifolia* litter to a level of some 20% higher than at the start of the experiment. An average increase of 10% was recorded in decomposing *Typha angustifolia* and *Phragmites communis* in Alderfen Broad. Similar increases in nitrogen occur in tree leaves decomposing in water and on land and are due to the activity of micro-organisms which fix nitrogen from the atmosphere (or water) while utilizing the carbohydrate source of the litter. This fixed nitrogen may be incorporated into different amino acids than were found in the original plant tissue, for instance proline occurs in greater concentration in well-decomposed *Phragmites* litter than in fresh litter. The increased nitrogen levels undoubtedly improve the quality of the litter as a food for other organisms.

The process of weathering is also an important cause of weight loss in decomposing material, though it is perhaps more difficult to isolate from other agents of breakdown and so has received little individual study. ANDERSON (1973) showed that in the autumn the leaves of sweet chestnut (*Castanea sativa*) on the surface of the litter remained dry. The action of the wind on the dry leaves caused them to crack and split along the veins and midribs. Eighty-six per cent of the weight of chestnut leaves was lost in a year and this was largely due to physical fragmentation.

3.5   Breakdown rates of litter

The rate of breakdown of litter can be studied by the paired plots method, by enclosing litter within bags, by tethering individual leaves or by using radio-isotopes. In the paired plots method the dead material from the first plot is collected and weighed at the beginning of the time interval, while the living material is removed from the second plot to prevent an increase of dead material on it. After a known interval of time the dead material in the second plot is removed and weighed. The difference in weight between the two plots, divided by the time, will give the breakdown rate of the litter. This method is unsuitable for sites where the litter is uneven, or blown about by the wind, but it has been used with some success in closed, even stands of vegetation, such as grasslands.

In another method, litter is enclosed within a bag made of suitable material, such as nylon, terylene or wire mesh. The mesh size can be varied to allow access to different types of organism (see Chapter 5), though for overall breakdown rates a mesh size large enough to allow the entry of animal decomposers, such as earthworms, is required. A lady's hair-net makes a perfectly adequate litter bag and experiments are easy to perform to compare the decomposition rate of different litter species, or of litter in different habitats (the breakdown of leaves on land and in water, for instance, can be compared). Leaves can be placed either whole within the bags or cut into discs of about 1 cm diameter. If a number of replicates are placed in position together, batches can be removed at regular intervals to study the breakdown rate of litter over a period of time.

In a third method, individual leaves are tethered by nylon threads and can be positioned on the litter surface or deep within the litter layer, thus enabling the breakdown rate at different depths to be compared. The method has the advantage that the leaf is as near to natural conditions as is possible. ANDERSON (1973) showed that leaf material confined within coarse-mesh litter bags lost 10–15% less weight than tethered leaves, which he suggested was due to differences in the feeding activities of soil animals in confined and unconfined litter. Furthermore the fauna within the bags (throughout the period of exposure) resembled that which occurs in the superficial layers of litter, so that the weight loss recorded was that of material artificially maintained in the surface regions of litter and hence may not accurately reflect the natural pattern of breakdown. Despite this difficulty, however, the litter-bag technique probably has the greatest general usage, especially in comparative studies of breakdown rate.

Radio-isotopes have rarely been used in decomposition studies. SMITH (1966) labelled plants with ^{14}C and then ground the various plant components to a powder and incubated them in soil. He then measured the rate of evolution of $^{14}CO_2$ from the soil resulting from the activities of microorganisms and was able to compare the decomposition rates of roots, shoots, petioles, etc., of several plant species under standard conditions. The method only allows the breakdown of compounds containing carbon to be studied. Furthermore, if carbon dioxide is fixed by autotrophic organisms in the soil the decomposition rate will be underestimated.

The rate of breakdown of plant litter, especially of leaves in forest soils, has received a good deal of attention. Some of the results of these studies are given in Table 3, where the approximate half-life for decay (i.e. time for 50% of the weight to be lost) is shown. Access to the litter by micro- and macro-organisms was allowed in the studies listed. Breakdown rates obviously vary greatly from species to species. For instance birch and lime leaves decompose much more quickly than oak leaves in good soils, whereas the decomposition of beech leaves and pine needles is very slow. Twigs and roots decompose more slowly than leaves. The decomposition of tree leaves in water is somewhat quicker than on land, though oak is still more resistant to decay than willow or sycamore. The leaves of the two reedswamp species break down more slowly than tree leaves. The decay of animal matter has received little study, though the breakdown rate of the fiddler crab *Uca* is very much faster than any of the vegetable materials listed in Table 3.

A number of factors are responsible for this variation in breakdown rate between species and between habitats, and in different years. Temperature is obviously important as it affects the rate of metabolism of decomposer organisms. The rate of litter breakdown in tropical forest is very much greater than that in temperate areas and there is no standing litter layer in the tropics. Pine and spruce litter decomposed 40% faster in southern Finland compared with the north of the country (MIKOLA, 1960). As well as these large latitudinal differences in decay, seasonal and diurnal changes also occur. Breakdown rates at winter temperatures are only half those in the summer, and the evolution of carbon dioxide from the soil is lowest just before dawn and highest in mid-afternoon (WITKAMP, 1969), this index of microbial activity closely following the daily temperature cycle. Also important, and often linked to temperature, are variations in moisture and aeration in the litter. In drought conditions animals move deeper into the litter layer or become inactive, whereas excess water reduces the aeration of the litter, the breakdown rate being markedly reduced in waterlogged conditions.

The structure and chemical content of the litter also influences its rate of decay. In the initial stages of decomposition hard leaves, with a thick epidermis, break down more slowly than softer leaves. The hard leaves of *Typha* are more resistant to decay than the softer leaves of *Phragmites* and the sun leaves of trees decay more slowly than shade leaves. Tougher leaves are presumably more impervious to the attacks of fungi and they are certainly eaten less readily by animals.

The majority of decomposing litter is acidic and coniferous litter is more acidic than hardwood litter. Many potential decomposers become less active when the pH falls below 5.0. Litter with a higher base content appears to decompose faster than litter from which most bases have been leached. Decomposition is said to occur more rapidly in material with more nitrogen (i.e. low C : N ratio). As stated earlier, however, micro-organisms themselves increase the nitrogen content of litter so that the observed microbial activity may be the cause, rather than the result, of elevated nitrogen levels. Even differences in the nitrogen content of leaves at abscission may be a reflection of the saprophytic population which has already colonized the leaves.

Polyphenols within the plant material may also influence the decay rate. These compounds make up 5–15% of the dry weight of a plant and many are leached out from the litter to give tannins, which cause precipitation of proteins. Polyphenols make leaves less palatable to invertebrates and the more polyphenol a leaf contains, the longer it takes to break down. This relationship does not necessarily hold between

Table 3 The rate of breakdown of material in various habitats.

	Species	Material	Habitat	Time for 50% breakdown (days)*	Reference
Birch	*Betula verrucosa*	Leaves	Mull soil	110	BOCOCK and GILBERT, 1957
			Moder soil	350	
Lime	*Tilia cordata*	Leaves	Mull soil	165	
Oak	*Quercus petraea*	Leaves	Mull soil	351	ANDERSON, 1973
Chestnut	*Castanea sativa*	Leaves	Moder soil	220	
Beech	*Fagus sylvatica*	Leaves	Moder soil	700	MASON, 1970
Beech	*Fagus sylvatica*	Twigs	Mull soil	1000	KENDRICK, 1959
Pine	*Pinus sylvestris*	Leaves	Mor soil	2500	
Grass	*Festuca arundinacea*	Roots	10 cm in soil	240	MALONE and REICHLE, 1973
Willow	*Salix* sp.	Leaves	River Thames	100	MATHEWS and KOWALCZEWSKI, 1969
Sycamore	*Acer pseudoplatanus*	Leaves	River Thames	140	
Oak	*Quercus robur*	Leaves	River Thames	210	
Reedmace	*Typha angustifolia*	Leaves	Eutrophic lake	370	MASON and BRYANT, 1975
Reed	*Phragmites communis*	Leaves	Eutrophic lake	210	
Cord-grass	*Spartina alterniflora*	Leaves	Intertidal saltmarsh	220	ODUM and DE LA CRUZ, 1967
Fiddler crab	*Uca pugnax*	Whole animal	Intertidal saltmarsh	15	

* Approximate times only, calculated from data in the papers referred to

species, however. ANDERSON (1973) found that sweet chestnut leaves were eaten preferentially to beech leaves, even when they contained a higher polyphenol content. However, it may be the concentrations of protocatechuic and gallic acids, rather than the overall polyphenol content, which affect the palatability of leaves and these two compounds are always present in lesser amounts in sweet chestnut than in beech.

A number of interacting factors therefore influence the breakdown rate of litter by influencing the activity of decomposer organisms. These organisms are discussed more fully in the following two chapters.

3.6　Models

The decay of vegetable material has been considered to follow an exponential model, i.e.

$$\frac{\mathrm{d}M}{\mathrm{d}t} = L - kM$$

or

$$M_t = M_0 e^{-kt}$$

where L is the rate of litter fall (mass per unit area); M_0 is the mass of litter per unit area at zero time; M_t is the mass of litter per unit area remaining after time t; k is a constant (the instantaneous decay rate) and e is the base of natural logarithms. If the litter mass is in a steady state, then

$$k = \frac{L}{M}$$

Equations of the exponential type have often been used to characterize the pattern of decay of different litters, but they cannot always be applied as not all materials decompose in this way. MINDERMAN (1968) considered that, while different chemical components of litter may decompose exponentially, the decomposition process as a whole could not be represented by a simple exponential function, but that the decomposition curve corresponded in shape with one constructed from the summation of a number of exponential functions. The decay of *Phragmites* litter can be shown to be essentially a linear process after the initial leaching of soluble materials in the first month. More comparative data on patterns of decay are required before an overall model of decomposition can be revealed.

4 The Role of the Microflora

4.1 Introduction

The saprophytic microflora involved in the decomposition of dead material includes the bacteria and fungi. The most important orders of bacteria which decompose plant litter are the Actinomycetales, Eubacterales, Myxobacterales and Pseudomales. Macro- and micro-fungi of many groups are saprophytes, the Basidiomycetes in terrestrial habitats and the Phycomycetes in aquatic habitats having received much attention. Fungal saprophytes have been discussed by HUDSON (1972).

Micro-organisms colonizing dead material may utilize the substrate in three main ways (BARLÖCHER and KENDRICK, 1974). They may directly decompose the substrate, using the energy obtained for growth. They may parasitize other organisms, or utilize their waste products, or they may use the substrate merely as an attachment site, while obtaining their nutritional requirements from the surroundings. This latter strategy is likely to be especially successful in aquatic habitats.

4.2 Succession

During the process of decomposition the substrate is continually changing, both physically and chemically, such that its suitability for colonization by different organisms also changes. An apparent succession therefore is seen. However the temporal pattern observed is based largely on the appearance of fruiting bodies, which are more readily identified, and it has been suggested that the succession is due to the time for each species to fruit rather than a succession depending on nutrition. The majority of studies have been of a descriptive nature and the functional role of individual species is still largely unknown. The following account describes successions recorded from various substrates in different habitats.

4.2.1 *Deciduous leaf litter*

Colonization of a leaf by bacteria and fungi occurs even before bud-burst so that a potential decomposing flora is present on a leaf throughout its life. The microflora on a living leaf is termed the *phylloplane* flora and persists for some time within the litter layer after leaf abscission before being replaced by a typical litter flora and eventually a soil flora.

HOGG and HUDSON (1966) followed changes in the micro-fungi on beech leaves from bud-burst until eighteen months after leaf-fall. They used three techniques to elucidate the succession. Direct observations were made to record the presence of fruiting bodies. Fungal spores were collected and examined by suspending leaf material over petri dishes containing a thin layer of potato dextrose agar. Cultural isolation of fungi was also done. Together, these methods enabled most of the microfungi to be characterized.

Three relatively distinct groups of fungi occurred on the beech leaves. Group 1 consisted of fungi which were present on the leaves before abscission and they persisted over the winter. The group contained *Discula quercina* (the conidial stage of *Gnomonia errabunda*), *Cladosporium herbarum* (the conidial state of *Mycosphaerella tassiana*), *Aureobasidium pullulans* (the conidial state of *Guignardia fagi*), *Alternaria tenuis* and *Botrytis cinerea*.

Group 2 appeared after leaf abscission and reached a peak the following spring, persisting until the end of the year. The fungi were *Discosia artocreas*, *Gnomonia errabunda*, *Mollisina acerina*, *Mycosphaerella punctiformis*, *M. tassiana* and *Guignardia fagi*.

Group 3 were recorded in the summer after leaf-fall, reached a peak in the following autumn and persisted until early spring of the third year of observations. More species occurred in this group than in the previous two: it consisted of *Polyscytalum fecundissimum*, *Spondylocladiopsis cupulicola*, *Microthyrium microscopicum*, *Mollisina* sp., *Lachnella villosa*, *Helotium caudatum*, *Endophragmia stemphylioides*, *E. catenulata*, *E. elliptica*, *E. laxa*, *Pistillaria pusilla*, *Chalara cylindrosperma* and *Doratomyces stemonitis*.

The frequency of occurrence of a representative of each of the three groups is shown in Fig. 4–1. Similar successions of fungi, but often with different species, have been observed from a number of types of deciduous and coniferous leaf litter.

Much less is known of bacterial successions on tree leaves. In the early stages of decay the species seem to be similar to those occurring in the phylloplane, though in greatly increased numbers. At later stages of decomposition species typical of the soil increase, especially the actinomycetes and spore-forming bacteria.

4.2.2   Bracken

FRANKLAND (1966) studied the succession of fungi on decomposing petioles of bracken (*Pteridium aquilinum*). She used three techniques (observation, anatomical investigation and isolation) and different fungi were cultured on bracken petioles to determine their role in the decomposition process. Bracken litter breaks down very slowly and the succession lasted over more than five years. Five groups of fungi were recorded throughout the period, but they were each dominant at different times (see Fig. 4–2). For example the Sphaeropsidales (e.g. *Phoma*) showed an early peak and the Phycomycetes (e.g. *Mucor hiemalis*) dominated later in the succession. The first colonizers, before and during senescence, were the weak parasites (e.g. *Rhophographus*) which formed

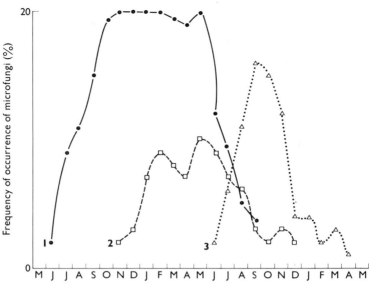

Fig. 4–1   The frequency of occurrence of three species of microfungi on monthly samples of twenty beech leaves. ● *Cladosporium herbarum,* ☐ *Discocia artocreas,* △*\Polyscytalum fecundissimum.* Redrawn, with permission, from HOGG and HUDSON, 1966.)

lesions in the outer cortex and penetrated into the non-lignified cortex and phloem, resulting in a loss of soluble carbohydrates and minerals. The initial substrate was high in cellulose and lignin and the next colonizers, the primary saprophytes, included organisms capable of breaking down these complex substances. Lignified cell walls were attacked by Basidiomycetes (e.g. *Trichoderma*) and after the first year the phloem had also been entirely obliterated. During the second year Basidiomycetes (especially *Mycena galopus*) dominated and cellulose, holocelluose (and to a lesser extent lignin) were attacked. Secondary saprophytes appeared, living mainly on the sugars produced in the decomposition of more complex compounds. By the third year the xylem cells were full of hyphae and 50% of the xylem had disappeared by the fourth year. Soil fungi began colonizing during the fourth year. By the fifth year of succession the pH of the petioles had risen from pH 5.2 to pH 6.1, while the carbon : nitrogen ratio had declined from 200 to 30.

4.2.3   *Phragmites*

TALIGOOLA *et al.* (1972) studied both the aerial and aquatic colonization of *Phragmites* leaves by fungi, using techniques similar to those described

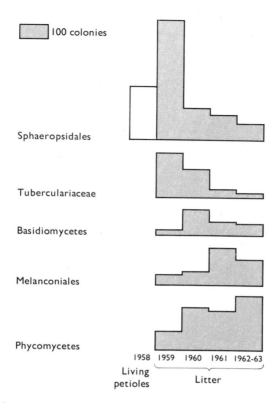

100 colonies

Sphaeropsidales

Tuberculariaceae

Basidiomycetes

Melanconiales

Phycomycetes

1958 1959 1960 1961 1962-63

Living
petioles   Litter

Fig. 4–2   The succession of fungal groups on washed particles of bracken (*Pteridium*). (Reproduced, with permission, from FRANKLAND, 1966.)

above. Colonization occurred in the very young leaves and the community was dominated by Hyphomycetes. Typical early colonizers were *Cephalosporium*, *Cladosporium* and *Alternaria*. As the leaves senesced in autumn other species of true saprophytes appeared, including *Epicoccum*, *Fusarium* and *Phoma*. Later colonizers of the moribund tissue included *Oidiodendron* and *Penicillium*, while at the late stages basidiomycetes such as *Trichoderma* occurred. As the reed stems became weakened by microbial attacks they collapsed and fell into the water. The aerial mycoflora flourished for a time if the stems floated but they eventually disappeared, to be replaced by an aquatic mycoflora. *Diplodina* is important in the early aquatic stage, to be followed by Ascomycetes, such as *Leptosphaeria* and Hyphomycetes, including *Articulospora*.

Large bacterial populations also develop on submerged leaves of reed and reedmace, dominants being *Pseudomonas fluorescens* and *Erwinia*.

4.2.4   *Laminaria fronds*

Successions of micro-organisms have received much less study in marine than in terrestrial or freshwater habitats. LAYCOCK (1974) has examined the bacteria associated with the surface of *Laminaria* fronds in Canada. He took samples from the tips, middles and bases of fronds. In the laboratory small pieces were cut from the samples and blended to produce a fine suspension of even consistency. Bacterial populations were then determined using a serial dilution technique and further characterization of the flora was done using isolation methods. The bacterial population was dominated by non-pigmented, oxidase-positive forms and these were divided into three groups, depending on the way glucose was metabolized. Glucose oxidizing forms were classified as *Pseudomonas* I and II and glucose inert forms such as *Pseudomonas* III and IV. There were also glucose fermenting *Vibrios*. Yellow-pigmented flavobacteria and tan-pigmented bacteria were also present.

Although no succession was recorded in the same way as fungi there were nevertheless marked seasonal variations in the groups occurring. During the summer there was a change from a glucose-oxidising (*Pseudomonas* I and II) to a glucose-fermenting (*Vibrio*) flora, which first started in the decomposing tips of the fronds in July, proceeding along to the meristem by September. The *Vibrio* flora was maintained throughout the winter, but in March the glucose-oxidizers again became important. Glucose-inert pseudomonads and flavobacteria maintained constant populations throughout the year, while the tan-pigmented group was more in evidence from August to November. The winter flora was dominated by psychrophilic forms, the summer flora by mesophilic forms. The psychrophilic bacteria were the only ones capable of hydrolyzing the carbohydrate laminarin and were isolated only when the temperature was below 7°C. Organisms hydrolyzing mannitol, gelatin and alginate were isolated when temperatures were above 7°C. Proteolytic bacteria were most abundant in the summer, when senescence of the *Laminaria* was also greatest. The relative rates at which different substrates of the seaweed were utilized was therefore dependent on temperature and hence season, and a progressive succession of bacteria did not occur.

4.2.5   *Rabbit dung*

Animal dung provides a highly suitable habitat for micro-organisms. It is very rich in nitrogen, due to the activities of the gut microflora and microfauna and it also contains good amounts of soluble carbohydrates, which are normally scarce in other substrates. High concentrations of vitamins, minerals and growth factors are also present. The original plant structure has been largely broken down and the dung has a high moisture content. Some invertebrate faecal pellets, however, are very compact and resistant to decay.

The fungal succession on rabbit dung has been studied by HARPER and WEBSTER (1964). Fresh pellets, collected in the field, were incubated in moist conditions. The first colonizers, appearing after 2–3 days incubation, were Phycomycetes, such as *Mucor hiemalis* and *Pilaira anomala*. They started declining after 7–10 days incubation, though they persisted for about three weeks. Discomycetes began to appear after six days incubation, representatives being *Ascobolus stictoideus* and *Saccobolus violascens*. These disappeared after 3–4 weeks. The Pyrenomycetes, such as *Podospora minuta* and *Sordaria fimicola*, did not appear until after 9–12 days of incubation. Basidiomycetes, that is *Coprinus* species, did not occur until late in the succession.

The first fungi to appear in the dung succession, the Phycomycetes, have rapid spore germination times, and quickly growing mycelia. They utilize sugars and are unable to break down complex carbohydrates. Ascomycetes and Hyphomycetes are able to break down cellulose, while the Basidiomycetes are largely responsible for lignin decomposition late in the succession. The importance of Phycomycetes in the early stages is due to the presence of sugars in the partly digested substrate.

4.3   The breakdown of wood

Most of the breakdown of wood is due to the Basidiomycetes of the families Agaricaceae, Polyporaceae and Thelephoraceae. The microorganisms can be placed in three major groups (KAÄRIK, 1974). The first includes fungi which live on cell contents but cannot decompose cell walls and consists of the moulds and blue-stain fungi. Blue-stain fungi (e.g. *Alternaria*) have pigmented hyphae which discolour wood.

Soft-rot fungi belong to the Ascomycetes and Fungi Imperfecti. Hardwoods are more vulnerable than softwoods to the many species of fungi involved. They are the first fungi to appear on newly exposed wood and flourish under moist conditions. The decomposition process is slow, with hyphae moving inwards from the surface and breaking down the cellulose of the cell walls, forming cavities within the wood. The decay process occurs only in the immediate neighbourhood of the fungal hyphae.

The third group consists of the *brown rot* and *white rot* fungi, all Basidiomycetes. Brown rot fungi (e.g. *Polyporus*, *Serpula*) attack structural polysaccharides, but not lignin, the action being diffuse throughout the cell wall. The attack is patchy and the framework of the cells appears to remain intact until it finally collapses.

Some white-rot fungi (e.g. *Fomes annosus*) attack lignins and hemicelluloses first, leaving cellulose until later, while other species (e.g. *Polystichus versicolor*) attack all the components of the cell wall together.

The secondary cell walls of the wood are gradually thinned from the lumen outwards, the decomposition occurring around the fungal hyphae. The attack is uniform rather than patchy.

If the dead wood remains on the tree, fungal succession can be very slow. The first Basidiomycetes, for instance, did not appear on an aspen branch until nine years after death.

4.4   Populations and biomass of micro-organisms

Dilution techniques and direct counts are frequently used for the estimation of populations of micro-organisms, but the results are often difficult to interpret. The organisms which grow on a dilution plate may have been dormant on the natural substrate, thus playing no active part in decomposition. Furthermore micro-organisms tend to be aggregated around particles so it cannot be determined whether colony counts on dilution plates emanate from a single individual or a colony.

There are also problems in estimating biomass. The separation of living micro-organisms from detrital particles is often impossible when the particle sizes of each group are similar. One way of overcoming this is to use as an index of biomass a compound which is a characteristic component of living tissues, but which does not occur in non-living materials. Such a compound must of course occur in relatively uniform concentrations in living cells. Adenosine triphosphate has been used in this way to estimate microbial biomass. Very small amounts of ATP can be assayed by making use of the bioluminescence reaction in firefly tails, one photon of light being emitted for every molecule of ATP which is hydrolyzed.

Bacterial counts in the soil, determined by serial dilution, are of the order of 10^6–10^8 per gram of soil. Direct counts, using a fluorescence microscope, give populations much in excess of this. Poor soils, such as sand dunes and podzols, tend to have lower bacterial counts than good agricultural soils. Bacterial counts on leaf litter are in the region of 10^9–10^{10} per gram of leaf. Biomasses of bacteria in agricultural soils range from 1600 to 35 000 kg/ha, though the estimates are very approximate. Fungal counts on decaying leaves are usually much lower than those of bacteria, 10^6–10^7 per gram of organic matter.

Similar counts are obtained from aquatic habitats. In a stream KAUSHIK and HYNES (1971) recorded bacterial densities of 10^4–10^7 per cm^2 on tree leaf litter of various species from January to May. Over the same period there were 10^2–10^4 fungal colonies per cm^2. On *Phragmites* litter in a shallow lake, bacteria averaged 8×10^5 per cm^2 after one month's exposure, whereas fungal colonies averaged 4×10^3 per cm^2. LAYCOCK (1974) recorded bacterial densities of 10^3–10^5 per cm^2, depending on season, on *Laminaria* fronds in the sea.

4.5   Interactions between micro-organisms

As discussed in section 4.2 saprophytic micro-organisms obtain their nutrients from the dead tissues of plants or animals, or from the waste products of the latter. They may also obtain their food from the soluble products of other micro-organisms or from the excretory products of higher plants. Micro-organisms associated with the excretory products of roots make up the *rhizosphere* flora. As well as producing soluble materials utilizable by other species, bacteria and fungi are also fed upon extensively by animals (see Chapter 5). An excellent review of the role of fungi in ecosystems is that by HARLEY (1972).

Fungi produce external enzymes which break down substrates to compounds of smaller molecular weight, which can be absorbed through the hyphal walls. Two kinds of enzymes are produced, those which are released and diffuse into the substrate and those which remain on the hyphal surface. Such external enzymes cannot be controlled in the same way as internal enzymes, with the result that excess breakdown products may be formed. WALSH and HARLEY (1962) studied the invertase activity of *Chaetomium globosum*. The invertase remains attached to the mycelial surface and hydrolyzes sucrose to glucose and fructose. Fig. 4–3 shows

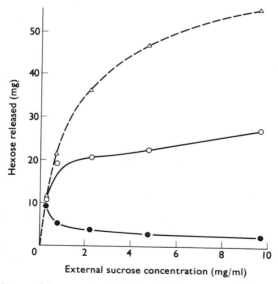

Fig. 4–3   Release of hexoses by hydrolysis of sucrose and the quantities of glucose and fructose absorbed by *Chaetomium globosum* as effected by sucrose concentration. Δ, total glucose and fructose released by sucrose breakdown; o, glucose uptake; •, fructose uptake. (Reproduced, with permission, from HARLEY, 1972.)

that as the concentration of sucrose increases, more glucose and fructose are produced than are actually absorbed, leaving a surplus in the external environment Glucose is absorbed preferentially by *Chaetomium*, causing an excess of fructose to be left. In this way it can be seen how micro-organisms with enzymes capable of breaking down complex molecules can alter substrates in such a way that other forms (e.g. sugar fungi), which can utilize only simple compounds, can colonize them.

Micro-organisms can release nutrients but they may also release compounds which inhibit the growth of other species. Most observations on this subject have been made in the laboratory. Competition occurs both for nutrients and for specific substrate sites, the amount of interaction depending on the nutrient concentration, and being modified by factors such as pH or temperature. Soil bacteria appear to be especially involved in antagonistic interactions with fungi, and a concentration of bacteria around a hypha will often lead to the disintegration or *lysis* of the hypha. It is also found that in many soils, especially fertile ones, fungal spores are inhibited from germinating. These general antagonistic effects are known as *mycostasis* or *fungistasis*. It has been shown that *Penicillium*, if it is the first to become established on decaying wood, will inhibit the lignin-destroying fungi, while *Trichoderma* produces antibiotics which cause hyphal lysis in other species. The overall importance of these interactions in natural ecosystems, however, is still open to question and it would seem likely that nutrient availability is more important in determining the type of microbial community which develops.

Very few attempts have been made to determine the relative importance of bacteria and fungi in decomposition. An obvious way to do this is to selectively inhibit groups of micro-organisms and observe the effect on the decay process. However, these techniques may stimulate the activity of the unaffected groups by enabling them to feed on the dead remains of the inhibited forms or by releasing them from competition. Despite these difficulties, however, the results may prove instructive.

KAUSHIK and HYNES (1971) studied the relative importance of bacteria and fungi in decomposing elm leaves in streams. They placed one group of leaf discs in stream water containing antibacterial agents (streptomycin and penicillin), another group in water containing antifungal agents (nystatin and actidione), a third group in water containing both types of agents, while a fourth group (in stream water only) acted as controls. The experiment ran for four weeks. The greatest weight loss in the leaf discs occurred in the controls, the least in the dish containing both types of antibiotic. The loss in weight was very small in the dishes containing antifungal agent, but the loss in dishes with the antibacterial agent was only slightly less than the control. Thus fungi appeared to be more important than bacteria in decomposition.

A similar experiment, using discs of dead *Phragmites* leaves in water from Alderfen Broad, Norfolk, was done to see if fungi were equally important in a lake. The same types of antibiotics were used. The weight loss of the material was measured after 33 days and the respiration rate of the litter was determined in a Warburg respirometer after 35 days and 122 days. The percentage contributions of fungi and bacteria to total respiration and weight loss of reed litter are shown in Table 4. Fungi were

Table 4 The percentage contributions of fungi and bacteria to total respiration and weight loss of *Phragmites* litter.

	Fungi	Bacteria	Fungi + bacteria
Respiration of litter			
after 35 days	31.4	16.3	62.9
after 122 days	4.1	45.1	49.4
Weight loss of litter			
after 33 days	24.9	24.6	53.0

important in the early stages of decomposition (though not statistically significantly more than bacteria), but after 122 days, fungi contributed very little to community respiration, whereas the contribution of bacteria was large. Organisms not affected by antibiotics were also responsible for much respiration. Serial dilutions of homogenized reed tissue after 35 days showed that very few fungal colonies survived in the mixture of antibacterial and antifungal antibiotics, while the population of bacteria in this mixture was only 4% of that in the antifungal solution. This residual respiration was probably largely caused by the microfauna and possibly by epiphytic algae. The experiment as a whole implied that fungi became less important in the later stages of decomposition.

The distributions of fungi and bacteria are different in the early stages of leaf decomposition in freshwater. SUBERKROPP and KLUG (1974) used a scanning electron microscope to show that the fruiting bodies of hyphomycetes (the dominant fungal group) occurred on the surface of the leaf material, with the vegetative hyphae and hence most of the biomass growing within the leaf tissue. Bacteria were mainly confined to the leaf surface, though they also penetrated via the stomata. As the leaf cuticle was gradually removed bacteria were able to act on the tissue underneath. SUBERKROPP and KLUG showed that one hyphomycete, *Alatospora acuminata*, was dominant in the autumn, winter and spring when the temperature was below 15°C, while *Lunulospora curvula* dominated in the summer at temperatures above 18°C. This seasonal change in dominance resembles the situation described earlier in the chapter for bacteria decomposing *Laminaria*.

In terms of weight loss of material, the microflora are probably responsible for about 90% of decomposition in both terrestrial and aquatic habitats. The fauna, however, play an essential part in the decomposition process and these are discussed in the next chapter.

4.6   Applied aspects

Micro-organisms are greatly involved in the decomposition of the enormous amount of waste material produced by the human population, especially the breakdown of sewage and refuse.

The treatment of sewage can be divided into three processes, primary, secondary and tertiary treatment. In the *primary treatment* screening devices and settling tanks are used to remove much of the solid materials, yielding primary sludge, which amounts to as much as 50% of the total solids. *Secondary treatment* involves the oxidation of dissolved and colloidal organics in the presence of micro-organisms and other decomposers. The aerated conditions are obtained in *trickling filters* or *activated sludge tanks*. Trickling filters are beds of stone over which sweep slowly rotating arms, distributing waste intermittently on to the surface of the bed. Beneath the filter is a drainage system to remove effluents. The stones have a large surface area for microfloral attachment and they allow the ready passage of air and water. A secondary *humus sludge* collects at the bottom of the filter and waste water is often recirculated to increase the efficiency of the process. As well as a rich microflora, trickling filters have a varied community of invertebrates, including protozoans, nematodes, rotifers, oligochaetes and dipteran larvae.

In the activated sludge process a continuous flow of water is maintained through long channels, material being kept in suspension by air or by mechanical agitation. The wastes are passed into a secondary settling tank, most of the settled sludge then being recycled through the aeration tank. The microflora is dominated by bacteria and the fauna consists entirely of small forms, such as protozoans and nematodes.

The sludge produced by primary and secondary treatment is passed into sludge digestion tanks, where it is decomposed anaerobically by micro-organisms. The tanks are normally heated to a temperature of 27–35°C, methane gas produced in the process being used to heat the tanks. The supernatant liquor is usually reprocessed through secondary treatment tanks and the final digested sludge is less than one third the volume of the original.

The effluents which are released from the sewage works are, of course, still rich in minerals, particularly nitrates and phosphates, and these have been responsible for many eutrophication problems in rivers and lakes. *Tertiary treatment*, to remove specific problem compounds (especially phosphorus) from effluents, is being increasingly practised, though it is often very costly.

Domestic refuse has been largely disposed of in the past by tipping, but the rapid increase in quantity has made this very wasteful of land, as well as of potentially reclaimable materials. Incinerators and composting plants are being increasingly employed. The former are cheaper and quicker, but again wasteful. Composting is initially expensive in outlay but produces material of value as an agricultural fertilizer, as well as a top-soil in reclamation projects. With the soaring costs of fertilizers, compost produced from refuse may become increasingly competitive. Modern plants often combine the products of domestic refuse and sewage processing.

Incoming refuse is initially screened to remove large objects and materials which are readily salvageable. It is then separated, to remove as much of the metals, glass and plastics as possible as these would reduce the quality of the final compost. Shredding produces smaller particles (1–8 cm) to increase the surface area available for microbial attack. In the next stage the moisture level is adjusted to provide optimum conditions for micro-organisms and nitrogen and phosphorus are added as nutrient supplies. The addition of sewage sludge here will provide these nutrients. The mixture is then passed into fermentation tanks through which a controlled air-flow is blown, the process requiring adequate aeration. Agitation also accelerates the decomposition process. The early stages of decomposition are by mesophilic micro-organisms and their intensive activity causes temperatures to rise. Above 40°C the mesophilic flora is replaced by a thermophilic one and the pH rises. At 60°C the thermophilic fungi die, leaving bacteria and actinomycetes. When the easily decomposable materials have been used up, the activity decreases and the temperature falls, allowing recolonization by thermophilic fungi. These stages take only a few days but the making of the compost, in which condensation and polymerization occur to form the final humus, takes several months, involving a mesophilic flora at ambient temperatures. Maturity may be achieved within a much shorter time in industrial plants. The humus can then be further screened and sifted, depending on the quality of the final product that is required.

4.7   Fossil fuels

These have been formed by the arrested and incomplete decomposition of vegetable matter due to anaerobic conditions. Peat is formed in poorly drained areas of land with abundant plant growth (e.g. *Sphagnum* bog, *Phragmites* swamps) when bacterial action is retarded. The springy, brown peat consists of easily recognizable plant remains. As the peat becomes buried by later deposits, it gradually compresses and slow chemical changes, especially the loss of volatile components, result in a gradual conversion into brown coal or lignite, and later humic coal. Present day lignite was formed tens of millions of years ago, while humic coal is hundreds of millions of years old. The black, hard rock of humic coal shows little evidence of its vegetable origin.

The rain of phytoplankton into the anaerobic depths of the oceans forms a fine sediment which is rich in organic material. Normal bacterial breakdown is again prevented and, by a complex series of chemical changes, the sediment may be converted eventually into mineral oil. With later sediment deposition, the oil tends to migrate into rocks with sufficient pore spaces to act as reservoirs, further movement being prevented by impermeable rocks. Oil wells result.

5 The Role of the Fauna

5.1 Introduction

In this chapter we are primarily concerned with litter, or its associated microflora, as food for animals. Litter and sediments, of course, also provide shelter for animals, both from adverse climatic conditions and from predators. Litter is both moister and subject to smaller temperature fluctuations than the ambient air.

Some animals also use the heat of decomposition to incubate their eggs. Several species of brush turkeys (Megapodiidae) from Australasia construct large mounds of leaf litter in which their eggs are laid. The male birds guard the mounds and periodically turn over the rotting vegetation to adjust the temperature. In this way the hens are relieved from the vulnerability to predators while spending long periods incubating eggs. Some snakes will also lay eggs in heaps of compost.

5.2 The animal communities

A great diversity of animals can be classed as detritivores or decomposers. Protozoans and nematodes are ubiquitous, occurring at densities of the order of 10^6–10^7 m^{-2}. Characteristic groups in terrestrial litters and soils are the mites (Acari), springtails (Collembola), fly larvae (Diptera), woodlice (Isopoda) and slugs and snails (Mollusca). Earthworms (Lumbricidae) are of special importance in temperate soils and termites (Isoptera) in the tropics. The ecology of soil-dwelling animals has been thoroughly reviewed by WALLWORK (1970). In freshwater habitats, tubificid worms, chironomid (Diptera) larvae, isopods and amphipods are important while in marine habitats the dominant macrodecomposers include polychaete worms, amphipods and molluscs (both gastropods and bivalves). The diversity of form of one group of detritivores, the dipteran larvae, is illustrated in Fig. 5–1.

KITAZAWA (1971) has examined the biomass and metabolism of a number of forest soils, ranging from sites with alpine coniferous shrub vegetation to tropical rainforest. Some of his results are given in Table 5. Over 90% of the metabolism in all sites was due to detritivores, the small remainder being predator metabolism. The biomass of animals was generally high in all the moist sites and the low value in the tropical rainforest was probably due to the presence of a thin litter layer, with no accumulation of organic matter. Metabolism is related to temperature, being lowest in the cooler alpine and temperate areas, and reaching a peak in subtropical forests, where both biomass and temperature are high.

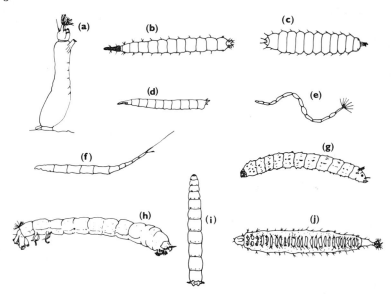

Fig. 5–1 Dipteran decomposers from various habitats. (a) *Simulium* (streams, filter feeders); (b) *Xylophagus* (rotting wood); (c) *Oxycera* (aquatic sediments); (d) *Sepsis* (dung); (e) *Ceratopogon* (aquatic sediments); (f) *Ptychoptera* (swamps); (g) *Bibio* (soils); (h) *Chironomus* (lake sediments); (i) *Tipula* (wet soils); (j) *Psychoda* (dung).

Table 5 Biomass and metabolism of the fauna of various forest soils. (Data from KITAZAWA, 1971.)

Forest type	Biomass (kJ/m²)	Respiratory metabolism (kJ/m²/y)
Alpine coniferous scrub	117	715
Temperate deciduous forest	109	694
Subtropical rainforest	109	2048
Tropical rainforest	38	865
Equatorial highland forest	380	1580
Equatorial highland dry forest	50	1141

The metabolism of detritivores in grasslands tends to be higher than in woodlands. MACFADYEN (1963), for instance, estimated a metabolism of 3177 kJ/m²/yr in a meadow grassland and 1840 kJ/m²/yr in a limestone grassland, his three temperate woodland sites having less than half of these values. The metabolism of the decomposer fauna of eutrophic lakes is in the same general range, of the order of 300–2000 kJ/m²/yr.

As well as moisture and temperature, the pH of the substrate has a marked effect on the structure of animal communities. Communities in mor humus, with a low pH, are dominated by mites and springtails, with few insect larvae, while earthworms and woodlice are scarce. Insect larvae become important in transition moders while in mulls, with neutral or alkaline pH, earthworms and woodlice are abundant, with mites and springtails less important. Snails, with their high calcium requirement for shell formation, are only common in alkaline soils.

Animals show preferences for specific depths in detritus which may be influenced by climatic conditions or food supply. In woodland litters, the oribatid mite *Carabodes labyrinthious* occurs in the surface layers, whereas *Nanhermannia elegantula* is mostly found in the AF layer, which consists of finely fragmented plant material (ANDERSON, 1971). In cold winter weather and in the hot, dry summer months oribatid mites tend to move into the deep layers of the litter, the degree of movement depending on the species resistance to desiccation. In the sediments of Loch Leven, the nematode *Monhystera paludicola* was most abundant in the upper layers of mud, while *M. filiformis* was found more often in deeper layers (BRYANT and LAYBOURN, 1964). Although food may be of influence, it appeared that *M. paludicola* had a higher sensitivity to a reduction in oxygen concentration.

Successions of fauna have been observed caused by changes in microclimate, the suitability of the food and by the colonizing ability of different species. DUNGER (1969) studied the development and succession of a decomposer fauna on re-afforested open-cast mining dumps. The trees planted were alder, poplar and false-acacia, with a shrub understory of lupins and melilot. After the first year of planting, conditions were dry and few animals were recorded. Two years later, in the shrub stage, conditions were still dry but a moder humus was beginning to form and fly larvae (Bibionidae) and springtails (*Hypogastrura*, *Tullbergia*) had increased markedly. Seven years after re-afforestation conditions had ameliorated and a mull-humus was developing, with the earthworm *Dendrobaena octaedra* abundant. After ten years the site had developed the appearance of woodland, and lumbricid worms, typical of mineral soils, had colonized it. The annual respiration of the animal community increased steadily over this ten year period, with arthropods accounting for the largest proportion after three years and worms after seven years. Dunger considered arthropods to be of greatest importance in the early stages of soil formation, and earthworms in the transition from moder to mull later on in the soil development.

The communities of animals which develop in dead wood will depend on the type of wood, its age and whether it is still attached to the tree or has fallen to the woodland floor. ELTON (1966) has described the bark beetles living in oak and ash. Four species of *Hylesinus* are commonly associated with ash, all occurring in different microhabitats. Four different species of beetle, again with different habitat preferences, live in elm wood. Their galleries ramify beneath the bark, enhancing the ability of microfungi to invade and providing microhabitats for other animals. Many of these, such as springtails, mites and fly larvae continue the decomposition process further. Other animals are predators, while a number use the galleries and crevices as resting places or hibernation sites.

Dead wood on the ground contains a far greater diversity of animals than that remaining in the canopy, probably because the latter is drier and subject to wider fluctuations in temperature. In one study of oak logs on the woodland floor, two hundred species of animals were recognized, the main association consisting of twenty-three species, of which mites and springtails were most abundant.

The colonization of wood by shipworms (Teredinidae), marine bivalve molluscs which bore into wooden structures, causes millions of pounds worth of damage annually to piers, sea defences, etc. Some shipworms utilize the wood primarily as food, while other species obtain most of their nutrition by filter feeding, the wood acting mainly as a shelter.

Dung and carrion provide highly specialized decomposer habitats, with very distinctive animal communities. Flies and beetles predominate in both. Over sixty species of insects were associated with carrion in one study and over a hundred species with dung in another. Scavengers, such as kites, vultures and jackals remove much carrion before it is colonized by micro-organisms and invertebrates and they still have a great role to play in maintaining hygiene in the poorer, hotter parts of the world.

5.3   Diets of detritivores

Decomposer animals are often considered to be unspecialized feeders, taking whatever is available. Superficially this is true and the guts of many species will contain the remains of several plant tissues, together with bacteria, fungal hyphae and spores, and the chitinous exoskeleton of arthropods. However, on closer inspection, subtle differences in diets between species can be detected. The age of the litter, the species of litter and its spatial position in the litter layer will all affect its palatability.

As an example of these differences in diet we can consider a study on woodland molluscs (MASON, 1970a). An examination of the faeces produced by several species of snails showed that they all ate mainly dead plant material, but there were distinct differences in the secondary foods taken (i.e. those eaten in smaller quantities). The faeces of *Arianta arbustorum* and *Hygromia striolata* contained chlorophyll, indicating that the snails had fed on living plants. The faeces of *Discus rotundatus* and *Marpessa laminata* contained rotting wood, the former also having large amounts of fungal hypae and spores. *Oxychilus alliarius* and *O. cellarius* had significantly more animal material in their faeces, showing that they also ate carrion. These observations were carried further by feeding various leaf litters to *Discus*. It was found that sycamore, sweet chestnut and oak litter were eaten in greater quantities than hornbeam and beech (the dominant litter type in the study area).

Swedish work has shown that different molluscs prefer different litter types. For example *Arianta arbustorum* prefers ash, *Cepaea hortensis* lime and *Eulota fruticum* birch. Studies on earthworms indicate a preference in the following order: elm > birch > ash > lime > oak > beech. When offered five species of tree leaves, elm was the most preferred by stream dwelling isopods and amphipods (see Fig. 5–2).

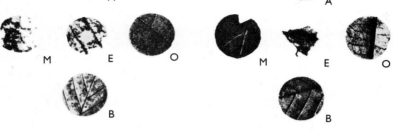

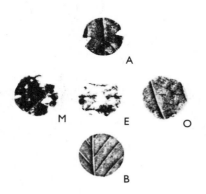

Fig. 5–2 Leaf preferences by *Hyalella* and *Gammarus* (amphipods) and *Asellus* (isopod). (A) Alder; (B) beech; (E) elm; (M) maple; (O) oak. (Reproduced, with permission, from KAUSHIK and HYNES, 1971.)

Differences in palatability between leaves may depend on morphological characteristics, such as the hairiness or thickness of the epidermis. Chemical differences are important and in particular the polyphenol content. Polyphenols gradually accumulate in leaves through the growing season. They cause precipitation of proteins in the leaves and may also interfere with the functioning of digestive enzymes. Beech and oak leaves have more polyphenols than others and are eaten less readily by animals. Within the litter layer polyphenols are gradually leached out or degraded by micro-organisms, making the leaves more palatable.

Experiments on the food preferences of decomposers are relatively easy to perform. Selected animals can be offered a simultaneous choice of leaf discs of a number of litter types in a stratified random way and the discs which are eaten first can be observed. Alternatively, discs can be offered one at a time and the amount eaten (either weight or area) in a given time (e.g. 48 h) can be measured. It would be interesting to compare the preferences of similar animals living in different habitats, for example the isopods *Oniscus asellus* from woodland and *Asellus aquaticus* from streams. To determine the preferences of the decomposer community as a whole, litter of various species can be weighed out into large mesh litter bags and placed in different habitats (e.g. woodland, meadow, stream), bags being collected and the remaining litter weighed at various intervals to compare overall weight loss of different species in various sites.

After the initial leaching period plant litter consists mainly of structural polysaccharides, which are difficult to break down, and many detritivores as a consequence have low assimilation efficiencies. Some examples are given in Table 6. The figures are only approximate, no account being taken of different foods. Furthermore there are often conflicting data (e.g. with mites) because experiments are difficult to perform, especially with smaller animals. The relatively low efficiencies are due to the absence of enzymes capable of breaking down structural polysaccharides. However the high efficiency shown in Table 6 for molluscs is because they do have polysaccharidases. Cellulases, pectinases, xylanases and chitinases have been found in the guts of a wide variety of molluscs, and also in some fly larvae and earthworms. Termites

Table 6 The assimilation efficiencies $\frac{C-F}{C} \times 100$ of some detritivores, where C is the weight of food consumed and F the weight of faeces produced.

Detritivore	Assimilation efficiency	Detritivore	Assimilation efficiency
Earthworms	1%	Mites	10–65%
Woodlice	15–30%	Molluscs	50–75%
Amphipods	15%	Molluscs (on carrion)	78–88%
Millipedes	6–15%		

are similarly endowed. The enzymes are produced both by the animals themselves and by their gut microflora. Note in Table 6 that when snails are feeding on carrion (in this case dead earthworm) their assimilation efficiency is similar to that of true carnivores.

5.4   Interactions with the microflora

The micro-organisms in the guts of detritivores are essential for the effective digestion of food. Large numbers of bacteria and fungal hyphae are also ingested incidentally when animals eat litter and this could effect decomposition rates by enhancing or suppressing the microflora. A careful examination of the feeding habits of animals in recent years has shown that some, e.g. certain collembolans and mites, feed exclusively on bacteria or fungi, litter in these cases being the incidental food. The aquatic snail *Bithynia tentaculata* can be maintained indefinitely on a suspension of bacteria. Three species of tubificid worms living in the mud of Lake Huron were found by BRINKHURST and CHUA (1969) to eat the sediment unselectively, but to utilize the bacterial component selectively and a different bacterium was found to survive passage through the gut in each species. Thus *Aeromonas* survived passage through the gut of *Peloscolex*, *Micrococcus* through *Limnodrilus* and *Pseudomonas* through *Tubifex*. Other bacteria were absent after digestion. This differential utilization of the bacterial resources may explain in part how many species of closely related tubificids can survive together in the same habitat.

Litter which is rich in micro-organisms is often a preferred food to litter which contains few bacteria and fungi. KAUSHIK and HYNES (1971) compared the feeding activity of the amphipod *Gammarus lacustris* on discs of elm leaves which had been incubated in antibiotics with control discs. Other discs were autoclaved to kill micro-organisms. Feeding on the control discs was much greater than on the autoclaved material or on discs treated with antibiotics.

Fungal and bacterial production is stimulated by detritivores despite the heavy grazing pressure. FENCHEL (1970) fed the marine amphipod *Parhyalella whelpleyi* on litter of the turtle grass *Thallassia testudinum*. After 92 hours there were few large-sized particles of detritus remaining due to the comminuting activity of the amphipods, while the detrital oxygen consumption was doubled. While bacteria were the main source of nutrition for *Parhyalella*, the increase in surface area of the detritus due to the amphipod enabled a greater population of bacteria to be maintained.

WELSH (1975) has examined the feeding of the grass shrimp *Palaemonetes pugio* on detritus of *Spartina* using a scanning electron microscope. The grass shrimp was abundant in saltmarshes and its production of faeces and dissolved organic matter outweighed its biomass production by 15:2. The shrimp fed by plucking away the cellular matrix from the surface of large detrital fragments, resulting in a varied assortment of uneaten particles which rapidly became colonized by micro-organisms. In this way decomposition of *Spartina* was accelerated.

In a lake, the amphipod *Hyalella azteca* was shown by HARGRAVE (1970) to ingest the bottom sediment randomly, but to assimilate selectively, taking only living material such as diatoms and bacteria. Nevertheless, the oxygen production (i.e. photosynthesis) of algae and oxygen consumption of bacteria was stimulated within a range of natural densities of the amphipod. However, if the density was increased above normal, the amphipods overgrazed their food and diatom populations decreased before bacteria due to their slower turnover time.

By increasing the surface area of detritus, animals improve its moisture-holding capacity, which also stimulates production of the microflora. Animals may further stimulate microfloral growth by grazing off senescent colonies, increasing nutrient turnover. They can aid fungal dispersal by carrying spores, the spread of Dutch elm disease by scolytid beetles being one example. Relationships between bark beetles and fungi are often symbiotic.

5.5   Animals and litter breakdown

Animals are important in litter breakdown in a number of ways. They eat and fragment large quantities of material, but digest very little and as described above this increases the surface area available for bacterial colonization. It also makes it easier for fungal hyphae to penetrate the tissue. The faeces of detritivores are also readily colonized by bacteria. The respiration rate of the faeces produced by an aquatic snail, *Lymnaea peregra*, was twice that of the detrital food. Earthworm casts contain many more bacteria than the surrounding soil.

The selective assimilation of nutrients by animals results in a material of different chemical (as well as physical) composition available to the microflora. Humic compounds formed as plant residues are gradually combined with the soil and with the mucus of animals. The animals also produce a homogeneous mixture of humus and soil by mixing and reworking the substrate.

Aquatic organisms can also alter the structure of the sediment. For example the midge larva *Tanytarsus* occurs at very high densities (over 100 000 per m²) in some lakes and lives in tubes of sediment, bound together with mucus. The aggregated particles in the tube persist for much longer than their inhabitants, increasing the overall size distribution of the sediment particles. In estuaries, deposit-feeding molluscs attain high densities and they continually rework the sediments. Heavier particles are gradually worked downwards leaving a fine, fluid surface silt, which is easily suspended in the water by slight currents. The gills of suspension-feeding molluscs are clogged by this sediment and larvae cannot establish. The activities of the deposit-feeders are here having a marked effect on the animal communities.

The importance of animals in the breakdown of litter has been examined in a number of ways. It is possible, for example, to study each detritivore in depth, looking at diets, assimilation, metabolism, etc, and summing results for all members of the decomposer community. Time, however, is usually prohibitive and a number of short-cut methods can be adopted. These involve exclusion experiments (keeping animals of selected sizes away from decomposing litter), or enclosures (keeping animals of selected sizes within decomposing litter), or selective killing experiments, using insecticides, vermicides or molluscicides to eliminate different animal groups from the litter.

EDWARDS and HEATH (1963) placed discs of oak and beech leaves into nylon mesh bags of different mesh-apertures to exclude various sizes of organism:

7 mm opening	all micro-organisms and invertebrates could enter
1 mm opening	earthworms excluded
0.5 mm opening	only micro-organisms and small invertebrates (e.g. mites, springtails) could enter
0.003 mm opening	all but micro-organisms excluded.

The bags were buried at a depth of 2.5 cm in cultivated pasture soil and were dug up every two months, the area of disc remaining being measured. In the 0.5 mm bags the breakdown of the litter discs was three times slower than in the 7 mm bags. The rate of breakdown in the 0.003 mm bags was even slower, the microflora having little visible effect in the absence of animals. The comminuting effect of animals would appear necessary to facilitate the invasion by bacteria and fungi in terrestrial soils.

Similar studies in freshwater (MASON and BRYANT, 1975), using discs of the relatively hard leaves of reed (*Phragmites*) and reed-mace (*Typha*) in bags of two mesh-sizes (see Fig. 5–3), showed that animals caused relatively little decrease in weight of the discs despite being present in large numbers. Combining these observations with laboratory experiments it was suggested that the fauna may be responsible for about 10% of the total weight loss of material.

In mull soils of temperate regions earthworms are particularly important in litter breakdown. They can have large populations (up to 500 per m²) and biomass (up to 250 g/m²). LOFTY (1974) has shown that the weight of leaf material that is buried in apple orchards is directly proportional to the weight of earthworms present. In an orchard with a high earthworm population the soil had a well developed crumb structure and no accumulation of litter on the surface. In an adjacent orchard, where repeated spraying with a copper-based fungicide

6 Decomposition and the Functioning of Ecosystems

6.1 Introduction

The process of decomposition can be summarized as follows. At death an initial period of rapid leaching occurs and populations of micro-organisms increase. Enzymic action breaks down some components of the litter, while the overall nitrogen content increases. Animals invade and either eat the litter, fragmenting it into smaller particles, or graze on the bacteria and fungi. The resulting faeces are rapidly recolonized by micro-organisms. There is a gradual decrease in particle size and litter biomass, with a simplification of chemical structure. The process is illustrated in Fig. 6–1.

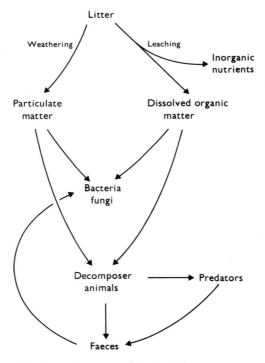

Fig. 6–1 Summary of the breakdown process.

6.2   Community metabolism

Many attempts to determine the total activity of the decomposer system have been made. Samples of soil or aquatic sediments can be studied under carefully defined conditions in the laboratory, for example in Warburg respirometers, but it is not known how the disturbance affects the respiration rates. Lake sediments can be collected in a relatively undisturbed manner using a standard corer and oxygen uptake from the water can then be monitored with an oxygen electrode.

Field measurements of respiration have been made for the soil/litter community. A typical method is that of WITKAMP (1966) who inverted circular boxes, with a surface area of 175 cm², into an otherwise undisturbed soil. The rims of the boxes were pushed 2.5 cm into the soil. After a period of an hour, in which the system was allowed to equilibrate, a dish containing 5 ml of 0.1 N KOH was placed in the box and the carbon dioxide evolution was measured at the end of the experiment by titration against hydrochloric acid. The carbon dioxide evolution from a woodland soil was estimated at 380 l/m²/yr by this method. A major drawback is that the total gas evolved will include that produced by the respiration of living roots and mycorrhizas, which in woodland soils may account for more than 50% of the total. The separation of the decomposition and production (root respiration) components of soil respiration is extremely difficult, but if we are to quantify and understand the major processes of ecosystems, the development of these short-cut methods will be essential.

6.3   Energy flow

The proportion of primary production which is utilized directly by herbivores or decomposers will vary in different ecosystems. In forest, scrub and desert ecosystems less than 10% of the annual primary production is grazed, whereas as much as 50% of primary production is accounted for by herbivores in savannas and managed grasslands. As much as 90% of the phytoplankton production of oceanic waters may be grazed. However, even here the faeces and corpses of zooplankton will eventually enter the decomposer system.

Streams and rivers flowing through forests have communities almost entirely supported by the input of materials from the land and this heterotrophic nutrition is typical of most running waters which have not been altered by man. A simplified diagram of a woodland stream ecosystem is shown in Fig. 6–2. The 'shredders' are those animals which feed on the leaf material directly, eating and breaking it into smaller particles. The 'collectors' are those which filter the water, taking out the comminuted particles and the faeces of the shredders. CUMMINS et al. (1973) investigated this community in experimental chambers, using the leaves

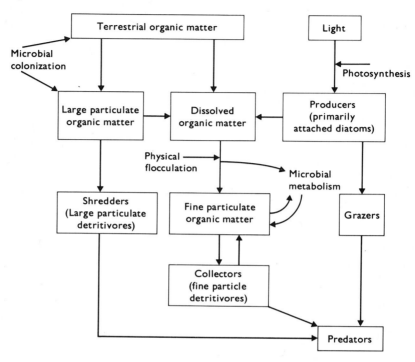

Fig. 6–2 The structure of a woodland stream ecosystem. (Reproduced with permission, from CUMMINS *et al.*, 1973.)

of hickory (*Carya glabra*) picked just before abscission and inoculated with hyphomycete fungi. After a period of seven days to allow fungal and bacterial growth, animals were introduced into the chambers in different combinations and at different densities. The shredders used were two species of cranefly larvae (*Tipula* spp.), three species of caddis fly larvae (*Pycnopsyche* spp.) and one species of stonefly (*Pteronarcys*). Four species of mayfly nymphs (*Stenonema*) were the collectors. The animals were allowed to feed on the hickory leaves for 40–60 days. At the end of the experiment, the mortality and growth of the animals was determined. The loss in weight of the leaves was measured and the proportions of dissolved and particulate organic matter were calculated.

The partitioning of the leaf litter biomass between the microflora, shredders and collectors in terms of growth and respiration is illustrated in Fig. 6–3. The growth of *Tipula* was found to be density dependent, being greatest at lower densities. Survival of *Tipula*, however, was higher in the presence of other shredders and lower when either alone or with the collectors, *Stenonema*. Shredders caused a significant decline in leaf weight. The growth of *Stenonema* was best at low densities, but with a high density of shredders. The collectors caused no significant decline in leaf weight, suggesting that when alone they fed on fine particles on leaf surfaces and their own faeces, rather than on leaf tissues. When the shredders were present, the collectors could feed on the particulate matter produced by these direct consumers and in nature the presence of shredders is obviously very important in the maintenance of collector populations.

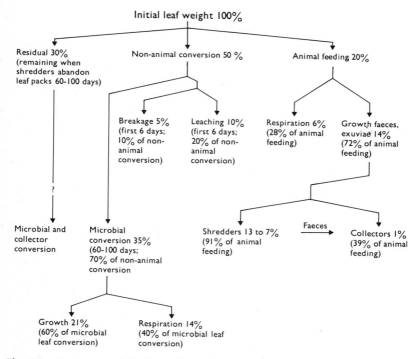

Fig. 6–3 Summary of the fate of a leaf during conversion to microbial and animal respiration and growth. (Reproduced, with permission, from CUMMINS *et al.*, 1973.)

The few completed studies of the energy flow through whole ecosystems have emphasized the importance of decomposition. An example is the work of TEAL (1962) on a salt-marsh in Georgia. The salt-marsh ecosystem is relatively simple because few organisms are able to withstand the great changes in salinity, exposure and temperature caused by the tidal cycle and the drainage from land. The producers are chiefly the grass *Spartina alterniflora* and attached algae, the herbivores are chiefly a leaf-hopper *Prokelisia* and a grasshopper *Orchelimum*. The mussel *Modiolus*, the snail *Littorina*, fiddler crabs *Uca* and *Sesarme*, annelids and nematodes are the chief detritus feeders. Spiders, dragonflies, birds and racoons make up the chief predators. A summary of the energetics is given in Table 8. Most of the energy fixed by the plants is used for plant

Table 8   The energies of a Georgia salt marsh. (From TEAL, 1962.)

Process	Energy $(kJ/m^2/yr)$
Input as light	2 508 000
Loss in photosynthesis	2 355 932
Gross production	152 068
Producer respiration	117 772
Net production	34 297
Bacteria respiration	16 260
Primary consumer respiration	2 491
Secondary consumer respiration	201
Total energy dissipation by consumers	18 952
Export from marsh	15 345

respiration, this high proportion probably being due to the energy required for osmotic regulation in the saline habitat. Herbivores assimilate very little of the net primary production (*ca.* 4%) most passing to the decomposers, of which the bacteria are the most important. Some 45% of the net production is exported as detritus to the estuary by the tides, where it is then decomposed further.

6.4   Nutrient cycling

The release of inorganic nutrients during decomposition is of profound importance in the maintenance of nutrient cycles and thereby plant growth. Some of these cycles in terrestrial ecosystems have been dealt with by JACKSON and RAW (1966).

An organism's role in the cycling of nutrients in an ecosystem may be very different from its role in energy flow. The estuarine mussel, *Modiolus*, for instance is responsible for about 0.8% of the total energy flow in a salt marsh (KUENZLER, 1961). The mussel feeds by filtering particulate matter out of the water and only a small proportion of this is utilized as food so that much of the filtrate is deposited on the mud surface as faeces and pseudofaeces, bound within mucus. Some 14% of the total phosphorus from the water is deposited in this way each day and made available to other salt-marsh organisms.

THOMAS (1969) examined the role of dogwood (*Cornus*), a shrub occurring in the understory of pine forests, in calcium cycling using the isotope ^{45}Ca. The dogwood, though having a much smaller biomass than the pine (only 0.2% of the forest biomass) was found to accumulate calcium disproportionately, especially in the leaves. At leaf fall this entered the litter layer, where it was released into the soil, the decomposition of dogwood leaves taking place at a much greater rate than those of pine. Thus dogwood, as with mussels, was far more significant in the recycling of nutrients than in the flow of energy.

A detailed example of nutrient cycling in ecosystems is furnished by the work of DUVIGNEAUD and DENAEYER-DESMET (1969) on two deciduous forests in Belgium. The mature oak–ash forest of Wavreille grew on deep, rich soil, derived from shale. Virelles was a mixed forest of oak, beech and hornbeam, about half the age of Wavreille and growing on shallow, calcareous soils. The results of the study, which involved collecting large amounts of data, using many techniques, are summarized in Table 9 and Fig. 6–4. Notice the very large pools of nutrients in the soils compared with those in the plant biomass. The soil weight and nutrient pool was much larger at Wavreille with its deeper, richer soils. An exception was for calcium, where the Virelles soil overlay calcareous bed-rock. The plant biomass on the mature Wavreille site was also much greater than at Virelles, but the net primary production (increase in biomass) was almost equal at both localities. The total uptake of nutrients, however, was different, calcium being taken up in greater amounts by the young forest, the other nutrients in greater amounts by the mature forest. These differences are, at least in part, due to the differences in the nutrient pools of the soils. Depending on the nutrient 60–80% of the annual nutrient uptake is returned to the forest soil, either through leaching or at litter fall, the remainder being retained in the vegetation as the biomass increment. The percentage retained by the younger forest of Virelles is somewhat greater than that of Wavreille. The annual budgets given in Table 9 and Fig. 6–4 obscure the seasonal processes of uptake, release and translocation which occur, as well as the differences in uptake and release by the different plant species and their litters.

Nutrients are cycling through the vegetation at two rates. There is the annual uptake of nutrients and their release back, via the litter and decomposition, to the soil pool. There is also the pool of nutrients retained within the plant biomass for long periods of time, to be released back to the soil after the death and decay of individual plants.

In tropical forests the high temperature and moisture result in a rapid breakdown of litter and release of nutrients. Furthermore the heavy rains cause a rapid leaching of nutrients out of the soil. In these conditions nutrient cycling is tight and the plants are very efficient at taking up nutrients. A greater proportion of the available nutrients is held within the plant biomass than is the case with temperate forests.

There is a general trend towards increasing litter layers with successional stages and the proportion of nutrients held within the biomass also gets larger. For example in the colonization of glacial soils, alder succeeds mosses and herbs after 10–15 years and the litter layer grows rapidly. There is a rapid increase in the nitrogen pools due to the nitrogen-fixing ability of the root nodules of alder, with a return of nitrogen to the soil via the litter (CROCKER and MAJOR, 1955). There is a decrease in the nitrogen pool in the soil when the climax vegetation of spruce and hemlock succeeds the alder. Alder also causes a decrease in the pH of the soil and this is true of a number of other species characteristic of intermediate stages of succession (e.g. bracken, heather, reeds).

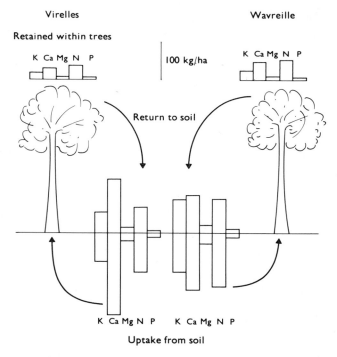

Fig. 6–4   Annual cycling of five inorganic nutrients in two Belgian deciduous woodlands (After DUVIGNEAUD and DENAEYER-DESMET, 1970.)

Table 9 A comparison of the production and nutrient cycling in two Belgian forests (after DUVIGNEAUD and DENAEYER-DESMET, 1969). Units are per hectare.

		Virelles	*Wavreille*
Total soil weight		1360 t	6318 t
Total plant biomass		156 t	380 t
Net primary production		14.4 t	14.3 t
Soil pools	K	26.8 t	160 t
	Ca	133 t	33 t
	Mg	6.5 t	50 t
	N	4.5 t	14 t
	P	0.9 t	2.2 t
Vegetation pool	K	342 kg	624 kg
	Ca	1248 kg	1648 kg
	Mg	102 kg	145 kg
	N	533 kg	1260 kg
	P	44 kg	95 kg
Annual uptake	K	69 kg	99 kg
	Ca	201 kg	129 kg
	Mg	19 kg	24 kg
	N	92 kg	123 kg
	P	6.9 kg	9.4 kg
Annual return to soil pool	K	53 kg	78 kg
	Ca	127 kg	87 kg
	Mg	13 kg	19 kg
	N	62 kg	79 kg
	P	4.7 kg	5.4 kg
Retained annually in increased biomass	K	16 kg	21 kg
	Ca	74 kg	42 kg
	Mg	5.6 kg	5 kg
	N	30 kg	44 kg
	P	2.2 kg	4 kg

6.5   Misuse of decomposition processes

The interference by man in the decomposition processes of ecosystems can result in rapid environmental degradation. In Chapter 4 the treatment of sewage was described. After treatment, the sewage is usually released into aquatic ecosystems as particulate matter. This will rapidly decompose further, a process requiring oxygen. In the laboratory the amount of oxygen utilized over a period of five days by a sample of water at 20°C is known as the *biochemical oxygen demand* (BOD) and is a measure of the decomposition activity of micro-organisms. If the BOD in a river is high, the concentration of oxygen within the water falls, resulting in the deaths of animals and plants. The composition of the fauna above and below a sewage outfall is very different. For instance, salmon and trout are replaced by coarse fish and even the most hardy of these are unable to survive gross organic pollution. The freshwater shrimp *Gammarus pulex* is unable to tolerate a reduction in oxygen and is replaced by the water louse *Asellus aquaticus*. With gross organic pollution only tubificid worms and a midge larva, *Chironomus thummi*, may survive and these occur in large numbers due to the rich food supply of organic matter and the absence of competitors and predators. The bottoms of organically polluted rivers are colonized by 'sewage fungus', a complex community of micro-organisms, dominated by a filamentous bacterium *Sphaerotilus natans*. As well as being aesthetically unpleasant, sewage fungus alters the benthic environment, making it unsuitable for other organisms.

The nutrients released by the decomposition of sewage may stimulate the growth of plants below an outfall, thus further unbalancing the natural ecosystem. Nitrates and phosphates are chiefly implicated and the problem is exacerbated where run-off of agricultural fertilizers is also occurring. Extensive growths of the blanket weed *Cladophora*, may smother the beds of streams, while in lakes large blooms of phyto-plankton develop. The changes in natural communities caused by nutrient enrichment are as profound as those due to oxygen depletion, but the effect is geographically much more extensive.

Many of our waterways which are used for sewage disposal must also be used downstream for drinking water supply. The nitrate released by the breakdown of sewage can present a health hazard to newborn babies through a condition known as *methaemoglobinaemia*. The nitrate is converted to nitrite by bacteria which thrive in the low pH conditions of the baby's stomach. The nitrites react with haemoglobin in red blood cells (the babies' haemoglobin being especially reactive) reducing the oxygen-carrying ability of the blood and causing asphyxiation.

Decomposition plays little part in modern, intensive agriculture. Most of the crop is removed and utilized. That which is left (e.g. stubble) is often burned rather than being ploughed back into the land and so many of the nutrients are dissipated to the atmosphere. The soil structure is often altered and large amounts of fertilizers are required to maintain the fertility. In the fourteen years between 1954 and 1968 the world consumption of fertilizer increased almost three times, the yield of crops increased somewhat less than this. This increase is disturbing in the case of phosphorus because, due to exceedingly slow global cycling times, it is effectively a non-renewable resource for man, though essential for maintaining productivity. A greater use of the processing of urban waste and sewage to produce a nutrient rich compost, as described in Chapter 4, may help to reduce the insatiable demand for inorganic fertilizers while solving some of the problems of water pollution. One of the difficulties of using man's wastes as compost is the frequent presence of large concentrations of heavy metals (e.g. lead, zinc, mercury) from industrial sources. These are often toxic to plants and present a health risk to man. However, as supplies of inorganic fertilizers become scarcer and more expensive, it will become economic to explore methods of removing these toxic materials to utilize the potentially valuable 'litter' of modern society in the decomposition cycle.

References

ANDERSON, J. M. (1971) *IV Colloquium pedobiologiae, Dijon (1970)*, 257–72 (mite distribution).

ANDERSON, J. M. (1973) *Oecologia*, **12**, 251–88 (tree litter breakdown).

ANDERSSON, F. (1970) *Bot. Notiser*, **123**, 8–51 (litter fall).

BÄRLOCHER, F. and KENDRICK, B. (1974) *J. Ecol.*, **62**, 761–91 (aquatic fungi).

BOCOCK, K. L. and GILBERT, O. J. W. (1957) *Pl. Soil*, **9**, 179–85 (breakdown rates).

BOLTON, P. J. and PHILLIPSON, J. (1976) *Oecologia*, **23**, 225–45 (earthworms).

BOYD, C. E. (1970) *Arch. Hydrobiol.*, **66**, 511–17 (reedswamp litter breakdown).

BRAY, J. R. and GORHAM, E. (1964) *Adv. Ecol. Res.*, **2**, 101–57 (forest litter production).

BRINKHURST, R. O. and CHUA, K. E. (1969) *J. Fish. Res. Bd. Canada*, **26**, 2659–68 (tubificid diets).

BRYANT, V. M. T. and LAYBOURN, J. E. M. (1964) *Proc. Roy. Soc. Edinburgh (B)*, **74**, 265–73 (nematode distribution).

CARLISLE, A., BROWN, A. H. F. and WHITE, E. J. (1966) *J. Ecol.*, **54**, 65–98 (litter fall).

COE, M. J. (1972) *East Afr. Wildlife J.*, **10**, 165–74 (elephant dung production).

CROCKER, R. L. and MAJOR, J. (1955) *J. Ecol.*, **43**, 427–48 (succession and nitrogen).

CUMMINS, K. W., PETERSON, R. C., HOWARD, F. O., WUYCHECK, J. C. and HOLT, V. I. (1973) *Ecology*, **54**, 336–45 (freshwater animals and litter breakdown).

DARNELL, R. M. (1967) in *Estuaries, AAAS Publ.*, **83**, 376–82 (decomposition in estuaries).

DUNGER, W. (1969) *Pedobiologia*, **9**, 366–71 (animal succession).

DUVIGNEAUD, P. and DENAEYER-DESMET, S. (1969) In *Analysis of Temperate Forest Ecosystems*, ed. D. E. Reichle. Springer-Verlag, New York, pp. 199–225 (nutrient cycles in forests).

EDWARDS, C. A. and HEATH, G. W. (1963) In *Soil Organisms*, ed. J. Doeksen and J. van der Drift). North Holland, Amsterdam, pp. 76–84 (litter bags).

ELTON, C. (1966) *The Pattern of Animal Communities*. Methuen, London.

FENCHEL, T. (1970) *Limnol. Oceanogr.*, **15**, 14–20 (marine amphipods).

FRANKLAND, J. C. (1966) *J. Ecol.*, **54**, 41–63 (bracken decomposition).

GRAY, K. R., SHERMAN, K. and BIDDLESTONE, A. T. (1971) *Process Biochemistry*, **6**, 32–6 (waste production).

HARGRAVE, B. T. (1970) *Limnol. Oceanogr.*, **15**, 21–30 (freshwater amphipods).

HARLEY, J. L. (1972) *J. Anim. Ecol.*, **41**, 1–16 (fungi in ecosystems).

HARPER, J. E. and WEBSTER, J. (1964) *Trans Br. mycol. Soc.*, **47**, 511–30 (fungal succession on dung).

HOGG, B. M. and HUDSON, H. J. (1966) *Trans Br. mycol. Soc.*, **49**, 185–92 (fungal succession on beech leaves).

HUDSON, H. J. (1972) *Fungal Saprophytism*. Studies in Biology no. 32. Edward Arnold, London.

JACKSON, R. M. and RAW, F. (1966) *Life in the Soil*. Studies in Biology no. 2. Edward Arnold, London.

JUNG, G. (1969) *Oecol. Pl.*, **4**, 195–210 (tropical litter fall).

KÄARIK, A. A. (1974) In *Biology of Plant Litter Decomposition*, ed. C. H. Dickinson and G. H. F. Pugh. Academic Press, London, pp. 129–74 (wood decomposition).

KAUSHIK, N. K. and HYNES, H. B. N. (1971) *Arch. Hydrobiol.*, **68**, 465–515 (decomposition in freshwater).

KENDRICK, W. B. (1959) *Can. J. Bot.*, **37**, 907–12 (breakdown rate).

KIRA, T. and SHIDEI, T. (1967) *Jap. J. Ecol.*, **17**, 70–87 (litter fall).

KITAZAWA, Y. (1971) In *Productivity of Forest Ecosystems*, ed. P. Duvigneaud. Unesco, Paris, pp. 485–98 (metabolism in forest litter).

KUENZLER, E. J. (1961) *Limnol. Oceanogr.*, **6**, 400–15 (mussels and phosphorus cycling).

LAYCOCK, R. A. (1974) *Mar. Biol.*, **25**, 223–31 (marine bacteria).

LEE, K. E. and WOOD, T. G. (1971) *Termites and Soils*. Academic Press, London.

LOFTY, J. R. (1974) In *Biology of Plant Litter Decomposition*, ed. C. H. Dickinson and G. H. F. Pugh. Academic Press, London, pp. 467–87.

MACFADYEN, A. (1963) In *Soil Organisms*, ed. J. Doeksen and J. van der Drift. North Holland, Amsterdam, pp. 3–17 (metabolism in grassland soils).

MALONE, C. R. and REICHLE, D. A. (1973) *Soil. Biol. Biochem.*, **5**, 429–39 (breakdown rates).

MANN, K. H. (1972) *Mar. Biol.*, **14**, 199–209 (decomposition in sea).

MASON, C. F. (1970a) *Oecologia*, **4**, 358–73 (mollusc diets).

MASON, C. F. (1970b) *Oecologia*, **5**, 215–39 (litter production).

MASON, C. F. and BRYANT, R. J. (1975) *J. Ecol.*, **63**, 71–95 (breakdown in freshwater).

MATHEWS, C. P. and KOWALCZEWSKI, A. (1969) *J. Ecol.*, **57**, 543–52 (breakdown in freshwater).

MELLANBY, K. (1972) *The Biology of Pollution*. Studies in biology no. 38. Edward Arnold, London.

MIKOLA, P. (1960) *Oikos*, **11**, 161–6 (litter fall).

MILLER, C. S. (1974) In *Biology of Plant Litter Decomposition*, ed. C. H. Dickinson and G. H. F. Pugh, Academic Press, London, pp. 105–28 (invertebrates).

MINDERMAN, G. (1968) *J. Ecol.*, **56**, 355–62 (decomposition models).

MORK, E. (1942) *Med. norske Skogfors-Ves*, **29**, 297–65 (litter fall).

NYE, P. H. (1961) *Pl. Soil*, **13**, 333–46 (tropical litter).

ODUM, E. P. and DE LA CRUZ, A. A. (1967) In *Estuaries, AAAS Publ.*, **83**, 383–8 (decomposition rates).

PHILLIPSON, J. (1966) *Ecological Energetics*. Studies in Biology no. 1. Edward Arnold, London.

PHILLIPSON, J., PUTMAN, R. J., STEEL, J. and WOODELL, S. R. J. (1975) *Oecologia*, **20**, 203–17 (litter fall).

RAPP, M. (1969) *Oekl. Pl.*, **4**, 377–410 (litter fall).

SMITH, J. H. (1966) In *The Use of Isotopes in Soil Organic Matter Studies*. Pergamon, Oxford, pp. 223–4.

STOATE, T. N. (1958) *Rep. For. Pep. W. Aust.*, **25**, 25 (litter fall).

SUBERKROPP, K. F. and KLUG, M. J. (1974) *Microbial Ecology*, **1**, 96–103 (micro-organisms).

TALIGOOLA, T. K., APINIS, A. E. and CHESTER, G. G. C. (1972) *Nova Hedw.*, **23**, 465–71 (fungi on reeds).

TEAL, J. M. (1962) *Ecology*, **43**, 614–24 (saltmarsh energetics).

THOMAS, W. A. (1969) *Ecol. monogr.*, **39**, 101–20 (dogwood and Ca cycling).

VIRO, P. J. (1955) *Comm. Inst. For. fennicae*, **45**, no. 6 (litter fall).

WALLWORK, J. A. (1970) *Ecology of Soil Animals*. McGraw-Hill, Maidenhead.

WALSH, J. H. and HARLEY, J. L. (1962) *New Phytol.*, **61**, 299–313 (activity of fungal hyphae).

WELSH, B. L. (1975) *Ecology*, **56**, 513–30 (feeding in shrimps).

WITKAMP, M. (1966) *Ecology*, **47**, 194–201 (soil metabolism).

WITKAMP, M. (1969) *Ecology*, **50**, 922–24 (soil metabolism).